Energy Conservation in Buildings and Industrial Plants

Energy Conservation in Buildings and Industrial Plants

Milton Meckler, P.E., C.Mfg.E.
President
Meckler Energy Group, Inc.
Encino, California

McGraw-Hill Book Company

New York St. Louis San Francisco Aukland
Bogotá Singapore Johannesburg London
Madrid Mexico Montreal New Delhi
Panama São Paulo Hamburg
Sydney Tokyo Paris

Library of Congress Cataloging in Publication Data
Meckler, Milton.
  Energy conservation in buildings and industrial plants.

  Includes index.
    1. Buildings—Energy conservation. 2. Factories—Energy conservation. I. Title.
TJ163.5.B84M42   1980   333.7   79-25041
ISBN 0-07-041195-6

Copyright © 1981 by McGraw-Hill, Inc. All rights reserved. Printed in the United States of America. No part of this publication may be reproduced, stored in a retrieval system, or transmitted, in any form or by any means, electronic, mechanical, photocopying, recording, or otherwise, without the prior written permission of the publisher.

1234567890   KPKP   89876543210

The editors for this book were Tyler G. Hicks and Susan Thomas, the designer was Mark E. Safran, and the production supervisor was Paul A. Malchow. It was set in Optima by Achorn Graphic Services, Inc.

Printed and bound by The Kingsport Press.

To Marlys, Ilyce, and Reneé
For all your patience and understanding

Contents

Preface   ix

One
Conservation in Exterior Design and Construction   1

Two
Energy Conservation Strategies in Interior Design   23

Three
Implementing Energy Conservation Measures   37

Four
Role of the Computer and Microprocessor in Energy Management   69

Five
Energy Conservation and Load Management   103

Six
On-Site Energy Systems   115

Seven
Industrial Sector Conservation Opportunities   157

Eight
Environmental Impacts of Energy Technology   197

Nine
U. S. Energy Policy at the Crossroads 219

Appendixes
A   Energy Conservation Checklist 237
B   Building Information for Total Energy
    Management 245
C   Conversion and Other Useful Tables 251

Glossary   257

Index   265

Preface

Energy conservation involves everyone. Our energy situation is currently worsening, and learning to deal with problems of limited energy supplies and escalating costs will be an ongoing process in the years to come.

This book was written to present a comprehensive view of energy conservation options in buildings and industry, not only from a technical standpoint but from a management perspective as well. Consequently, it contains useful information for the building owner and the construction and design engineer, for the plant engineer and operator, the industrial manager, and the architect. The principles outlined in this book can be applied equally by those who formulate national energy policy and by those who are responsible for maintaining large facilities, be they commercial, institutional, or industrial structures or processes.

The construction contractor will be exposed to new ways to consider the energy values of building materials and methods for reducing energy costs at the construction site.

The architect will gain insights into new design concepts usually evaluated by architectural consultants, which can enhance the potential of new structures to use energy wisely.

The design engineer will discover useful methods and information for planning mechanical and electrical systems that utilize energy more efficiently.

Building owners and managers will be familiarized with valuable techniques for identifying and quantifying energy consumption in their structures. They will also discover useful methods for cutting energy costs and for evaluating the economic advantages and disadvantages of new energy management systems.

Maintenance engineers will be sensitized to new, effective ways to operate their equipment efficiently and to motivate personnel to become more energy-conscious.

The industrial manager will gain insights into the predictive techniques needed for long-range energy planning.

In addition to this, the book contains a thorough analysis of energy usage in the iron and steel industry, in the pulp and paper industry, in the cement industry, in the chemical industry, and in petroleum refining. While dollar figures and fuel sources may change over the coming years, the basic approach contained herein will remain valid and useful for years to come.

In attempting to strike a balance between the technical and the management aspects of energy, we have selected a broad range of topics, including:

Conservation in exterior design and construction The initial planning of buildings and the energy intensiveness of construction materials are discussed.

Energy conservation strategies in interior design This section outlines useful methods for calculating energy usage within buildings.

Implementing energy conservation measures Under this heading a wide range of practical measures for reducing energy consumption is explained in detail.

The role of the computer and microprocessor in energy management Practical steps are outlined for selecting appropriate automated building systems.

Energy conservation and load management Concrete strategies are discussed for reducing the amount of energy consumed by HVAC systems and other machinery. Load-shedding techniques are outlined in detail.

On-site energy systems A wide range of options are outlined concerning the feasibility of on-site generation. Methods of determining the economic availability of such systems are explained in detail.

Industrial-sector conservation opportunities Problems of fuel substitution in industry are discussed. Opportunities and future possibilities for reducing energy consumption in all major industrial categories are outlined.

Environmental impacts of energy technology A discussion is presented of ways to reconcile growing energy needs with related environmental concerns and problems. Governmental constraints involving power generation are detailed.

U.S. energy policy at the crossroads This includes a timely explanation of the National Energy Act and its probable impacts. Anticipated trends for the future are also discussed.

In creating a book of this broad scope, we must naturally define certain boundaries; such a book cannot be all things to all people. Solar, nuclear, and various alternative energy sources, for instance, are dealt with from the managerial perspective rather than on a technical level. Likewise, the subject of retrofitting is not dealt with in great detail. Nevertheless, the principles and approaches outlined in this book are applicable to both new and remodeled building structures and engineered facilities. We have also provided references to give those interested in greater technical detail access to pertinent information.

The opportunities for developing effective energy conservation strategies are great. While there are few easy solutions, a concerted effort by us all may well bring about the major economic and technological advances we need for a secure future.

Milton Meckler

Energy Conservation in Buildings and Industrial Plants

Conservation in Exterior Design and Construction

While we will continue to need imported oil for many years, we should do everything possible to minimize its use.

Gerard C. Gambs

The energy crisis is real. Securing oil supplies has become a complex international problem, and there are few indications that the situation will change for the better in the foreseeable future. Other energy sources appear promising, but most of these new technologies are beset with difficulties of one kind or another. In any event, it will be many years before true alternative sources can be firmly established, and they will almost certainly be more expensive.

As supplies dwindle and costs soar, it becomes obvious that energy conservation is no longer optional. We must develop new and better ways to stretch our energy dollar. For economic as well as philosophical reasons, it is imperative that we move boldly to reduce waste.

1-1 TRANSITIONS IN CONSTRUCTION MATERIALS

It has been estimated[1] that by the year 2000, the United States will experience at least one materials crisis. Many of the commonly used construction materials of the seventies (i.e., lumber, steel, concrete) may have to be modified in form and use to meet changing needs.

Today, approximately one-third of our land surface is covered by forest. Of this estimated[1] 750 million acres (3.035142×10^{12} m²), approximately two-thirds is grown for construction needs. But U.S. Forest Service forecasts paint a gloomy picture for the U.S. production of softwood in the years to come; they predict a possible 12 billion bd·ft (3.66 billion bd·m) shortfall by the 1980s and a shortfall of up to 20 billion bd·ft (6.10 billion bd·m) by the year 2000.

Perhaps through better management of our forecasts, improved cultivation efforts, and the reduction or reuse of mill residues, these trends can be reversed.

U.S. Bureau of Mines estimates reveal a sufficient supply of iron ore to keep steel viable as a construction material for some time. Competition pressures may also bring about new advances. The American Iron and Steel Institute predicts steel with available strength of 300 kips/in² (2068 MN/m²) by the year 2000, although today it is available only in the 20 to 100 kips/in² range (137.9 to 689.5 MN/m²). This may seem awesome until we consider that even at 300 kips/in² (2068 MN/m²), we approach only 10 percent of the theoretical strength of steel.

Concrete should continue to serve the construction industries well in the next century. Major improvements in temperature, chemical resistance, and high-tensile, self-curving flexibility are anticipated. Except for the occasional scarcity of natural aggregates in some localities, problems of availability are not anticipated, although pricing problems may arise. Pyroprocessed shales or clays can also be substituted for natural aggregates where advantageous. Greater emphasis will be placed on manufactured light-weight aggregates exhibiting higher strength and greater durability. With selective impregnation or reinforcement, strengths of over 65 kips/in² (413.69 MN/m²) for special use and 20 kips/in² (137.9 MN/m²) for general use may be available. Expanding the operation of solid-waste incinerators may provide another economical new source of glassy residues which now appear promising as additives.

Despite an intensive effort in recent years to develop an active role for structural plastics, little impact has yet been made. This is in spite of such desirable properties as moisture and corrosion resistance, ease of maintenance, etc. Reasons for this disappointing performance include limited structural strength and lack of fire durability. While the use of exotic reinforcements can improve compression and tensile strengths appreciably, special fire detection and suppression systems may be required in buildings with plastic furnishings and structural members.

1-2 ENERGY INTENSITY OF BUILDING MATERIALS

Let us now examine the energy per pound needed to fabricate various building construction materials. Construction is itself a large consumer of energy, particularly when one considers the various manufacturing, assembly, and transporting steps required for each item of steel, copper, glass, concrete, aluminum, insulation, tile, carpeting, and so on.

Robert A. Kegel, P.E., a mechanical consultant, reported[2] using, on one of his firm's projects, a number of materials whose energy intensiveness is listed in Table 1-1. A three-story community college building, comprising approximately 432,000 ft² (40.133 m²) with a steel building structure enclosed by architectural steel and glass curtain walls, was built in Chicago, Illinois. It is interesting to note in Table 1-2 that

TABLE 1-1 Energy Intensiveness of Typical Building Materials

	\multicolumn{4}{c}{To Fabricate}
Material	Btu/lb	kJ/kg	Btu/unit	kJ/unit
Aluminum	41,000	19,636		
Ceiling materials	1,500	718		
Concrete	413	198		
Concrete blocks—8 × 8 × 16 in (20.3 × 20.3 × 40.6 cm)			15,200/block	7,279/block
Copper	40,000	19,157		
Drywall	2,160	1,034		
Glass	12,600	6,034		
Insulation:				
Duct—1 in (2.54 cm), 3-lb (1.36-kg) density			51,400/ft²	584,161/m²
Pipe—2 in (5 cm)			7,700/ft²	87,510/m²
Building (board)			2,040/ft²	23,184/m²
Paint	4,134	1,979		
Roofing			6,945/ft²	3,326/m²
Steel	13,800	6,609		
Vinyl tile	8,000	3,831		

NOTE: Many studies have been initiated by, among others, the Federal Energy Administration, to verify these types of data. At the time this research was conducted, much of this information was "classified" by the respective manufacturer. Some of those consulted in obtaining these data are listed here: (1) Aluminum Association of America, (2) American Brick Co., (3) Anaconda American Brass Co., (4) G.D.F. Corp., (5) Gustin Bacon, (6) Gold Bond Corp., (7) Pittsburgh Plate Glass Industries, Inc., (8) Portland Cement Association, (9) Republic Steel Corp., (10) U. S. Gypsum, (11) University of Illinois, Dept. of Forestry.
SOURCE: "Thermic Diode Solar Panels, A Brief Summary," *Proc. Int. Solar Energy Soc. Joint Conf.*, Winnepeg, Canada, **2**:1–23 (August 1976).

concrete used in the example building represents 58.7 percent of the total building weight but only 12.4 percent of the total energy used to fabricate all materials and equipment needed to construct the building. On the other hand, steel, which represents only 6.8 percent of the building's weight, consumes about 69.5 percent of the same total.

The distribution of energy used to fabricate the various materials and equipment employed in constructing this building can be found in Table 1-1. Material and equipment weights are also noted.

1-3 ENERGY USE AT THE CONSTRUCTION SITE

The highly competitive construction industry can be expected to react to the economic impact of energy costs and availability problems involving on-site construc-

TABLE 1-2 Distribution of Energy (E_m) Used to Fabricate Materials and Equipment Employed in Constructing Example Building

Material	% of Total Weight*	½E_m†
Aluminum	0.08	1.6
Ceiling materials	0.5	0.3
Concrete	58.7	12.4
Concrete block	27.0	1.9
Copper	0.2	4.3
Drywall	4.2	0.7
Glass	0.2	1.2
Insulation	0.04	4.9
Paint	0.04	0.1
Plumbing fixtures	0.2	2.0
Roofing	2.0	0.5
Steel	6.8	69.5
Vinyl	0.04	0.6
TOTAL	100.00	

* Total weight of materials and equipment = 240.0 tons (217.7 t).
† E_m = 476.190 Btu/ft² (5411.9 kJ/m²).
SOURCE: "Thermic Diode Solar Panels, A Brief Summary," *Proc. Int. Solar Energy Soc. Joint Conf., Winnipeg, Canada,* **2**:1–23 (August 1976).

tion operations. Fuel availability represents the primary concern since, without fuel to drive construction equipment, a shift to higher, labor-intensive (more costly) construction methods would be required. Because an increase in fuel costs affects competitive bids equally, the concern over fuel costs is generally of secondary importance. Recent studies[3] show, in fact, that the total impact of the energy component of total construction costs should not be used to materially influence the demand or cost of construction.

Refer to Fig. 1-1 for a plot of the energy consumed per $1000 of construction as a function of project size for various types of construction. Notice that the pattern is far from consistent. For example, the reduction of specific energy consumption is rarely in excess of 10 percent for a 10:1 increase in project size. Variations in consumption levels do vary significantly by construction type. Construction of highways and streets will consume three times as much energy per $1000 of construction cost as, say, single-family residences. Table 1-3 contains a percentage breakdown of energy sources by construction-project type. Energy costs as such for the years 1971 to 1974 amounted to approximately:

1. One percent of the construction cost for, say, residential and general building construction
2. Three percent for all heavy construction (excluding road building)
3. Seven percent for road building

Conservation in Exterior Design and Construction 5

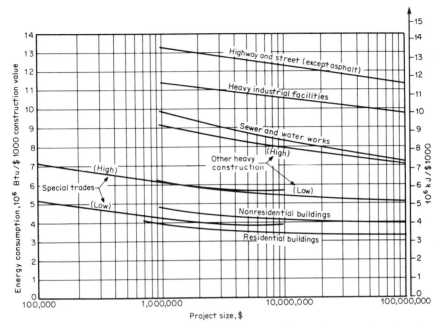

Fig. 1-1 Energy consumption per $1000 of construction and as a function of project size. ("Energy Use in the Contract Construction Industry," Federal Energy Administration, U.S. Department of Commerce, Report No. PB-245 422, February 18, 1975.)

Factors considered in analyzing the effect of climatic variations on these energy consumption totals included equipment warm-up periods and minimum space heating and cooling requirements. Climatic differences between Maine and Florida, for example, resulted in an energy consumption ratio of $8:1$.

After the oil embargo, most construction firms identified the operation and maintenance of construction equipment as an area for sizable improvement. For

TABLE 1-3 Percentages of Total Equivalent Btu Supplied by Major Sources

Source	Single-family Homes	Multifamily Homes	Nonresidential Buildings	Highways and Streets	Other Heavy Construction
Gasoline	64	63	65	9	70
Diesel	27	26	25	4	20
Lubricants	4	4	4	1	5
Purchased kWh	5	7	6	1	5
Asphalt/tars				85	
TOTAL	100	100	100	100	100

SOURCE: "Energy Use in the Contract Construction Industry," Prepared for the Federal Energy Administration, U. S. Department of Commerce Report no. PB-245-422, Feb. 18, 1975.

example, construction equipment five years old operates at markedly lower efficiencies than new equipment. Procedures to improve equipment efficiency and better construction equipment maintenance are needed to save scarce fuel. Delays are frequent on construction jobs, and personnel often leave vehicles and equipment idling. In addition, morning warm-up periods are often excessive. It is estimated[2] that proper enforcement of energy-conscious procedures and guidelines would avoid such abuses and reduce overall construction site fuel consumption by 75 to 85 percent.

Unfortunately, there are few opportunities for short-term substitution of traditional energy sources in construction. Limited opportunities, such as choices between diesel- or gasoline-powered equipment, do exist. It must be recognized, however, that smaller, less efficient gasoline-powered construction equipment may result in more fuel consumption than the larger, more efficient diesel-powered equipment.

Use of purchased power for on-site construction versus less efficient on-site portable electric power generation does not appear to offer significant advantages. Rigorous enforcement of energy-saving, equipment-use procedures will probably continue, since it is usually accompanied by a noticeable improvement in the overall productivity of associated labor and/or equipment. Improved equipment maintenance practices also appear promising because economic savings are readily identified and complications concerning negotiated labor practices are minimal.

In summary, because of its highly competitive nature, the construction industry is likely to address only the energy-savings practices which are clearly visible at the bottom line. All others will be passed on to its customers. Enforcement of energy-savings practices by regulatory agencies is difficult because of the fragmentation of the subcontracted work. It should be recognized that the long-term trends to substitute energy-consuming equipment for labor will continue despite present high and rising fuel costs. On the positive side, however, is a greater awareness when purchasing new construction equipment of the need to improve fuel economy. This is resulting in the upgrading of construction equipment efficiencies by manufacturers.

DESIGN CONSIDERATIONS FOR EXTERIOR BUILDING DESIGN

The building designer should take into consideration the proper building orientation with respect to the sun. Exterior shading devices, such as sun hoods and trees, can be used to control solar radiation. Air space and effective insulation materials in wall construction can control heat gains and losses from conductive transfer. Proper joint design and adequate weather stripping can help minimize heat gains and losses caused by the infiltration of outside air. The use of glass in new construction should be planned carefully. Passive options and passive-active options to reduce energy consumption should be considered. At the construction site, a reduction of morning warm-up periods and equipment idling time can reduce energy waste.

1-4 HEAT TRANSFER PRINCIPLES

To appreciate the problems of energy conservation and to find suitable solutions, the energy planner must have a firm understanding of the basic principles of heat transfer. One of the primary functions of a building is to act as an envelope which conserves heat when the outside is cold and which dissipates heat when the outside is warm. The envelope can do this because of the basic physical properties of building construction materials. In climates where it is generally warm or cold for the entire year, traditional materials and construction techniques function quite well without help from outside energy sources (the tropical hut and the arctic igloo). In regions where it can be both warm in the summer and cold in the winter, great care in the handling of materials and construction techniques is needed to allow interior comfort without the infusion of great quantities of outside energy in heating and cooling systems.

For buildings, the basic unit of heat measure is the British thermal unit (Btu) or watthour (wh). It represents the potential to raise the temperature of one pound of water (about one pint) at 60°F (15.55°C) by one degree Fahrenheit. It is about equal to the amount of heat given off in burning one ordinary wooden kitchen match. We commonly refer to heat used over time as a Btu per hour (watt) or Btu per year (watthour per year).

Once introduced into a building, heat is manipulated in the building envelope by three principles of nature: heat conduction, heat convection, and heat radiation:

1. Conduction is the transfer of heat from a hotter to a colder surface through direct contact. Thermal conductance C is the measure of this type of heat transfer and refers to the number of Btu per hour (watts) or Btu per year (watthours per year) that would flow through one square foot (one square meter) of a given material of a specified thickness to cause a temperature drop across the material of one degree. For composite constructions, the conductances of the various members are combined to give the overall gauge of thermal transmission, called the *U factor*. The lower the U factor, the more a particular construction resists heat passage. It is common, particularly for insulating materials, to give the inverse of the conductance to indicate the resistance to heat flow. This term is called the *resistance R*, with higher values of R indicating greater insulating value.

2. Convection is the transfer of heat from the mixing of various gases and liquids at different temperatures. This is not only a function of the temperature of the materials but also of the motion of the mixing fluids. There are two general types of convection in buildings: natural convection which relies on the action of gravity to set up the necessary motion and forced convection which relies on mechanical stirring to set up the motion. Forced convection can greatly increase the rate of heat transfer.

3. Radiation is the transfer of heat in the form of electromagnetic waves. This transfer occurs across a space and absorbs the waves between at least two bodies,

one the sender and one the receiver. The amount of transmission from a body is a physical property called its *emissivity*, which is dependent upon the nature and temperature of that body. The amount of transmission absorbed by a body is a physical property called its *absorptivity*, which is also dependent upon its nature and temperature and further dependent upon the particular wavelength of the incident radiation. Critical to this understanding is that solar radiation is shortwave radiation while radiation from earthbound heated objects is longwave radiation. Many construction materials behave very differently under the effects of shortwave and longwave radiation.

An additional property of heat transfer across building materials is the thermal mass of materials. Although the idea of thermal mass is quite complicated, it can be summarized that materials not only resist heat transfer but also tend to store heat energy within them for a fair period of time. For very dense materials that period of time may be from 5 to 10 hours, causing a temperature lag in the building throughout the day. This phenomenon, called *thermal lag*, may allow a building to store external heat in its walls during the day and then give it off to the building occupants over the course of the colder night, thus decreasing the need for interior heating at night or exterior cooling in the morning.

The building envelope is made up of many different mechanisms which operate under the principles explained above. Although there are too many of these to consider in detail, they can be grouped into five major categories of heat flow in and out of buildings:

1. Solar gain through windows
2. Heat gain from conductive transfer through all walls exposed to direct solar radiation
3. Heat gains and losses from conductive transfer through surfaces exposed to outside air temperatures
4. Heat gains and losses from infiltration of outside air through cracks and openings
5. Heat gains and losses from ventilation with outside air

Solar gain through windows depends upon the orientation of the windows to the sun and the characteristics of the glass. A horizontal skylight has a maximum solar exposure, but a north-facing window has almost none. The reflectivity of various types of glass controls, for a given window orientation, how much of the direct solar radiation will be passed into the building and how much will be reflected away. Exterior shading can mitigate such loads.

Heat gain from conductive transfer through surfaces exposed to direct solar radiation can be considerable. Important factors here are the type of material exposed, the color of the material, and the orientation of the materials to the sun. Radiation loads are most severe for the roof and the east-west walls during the

Conservation in Exterior Design and Construction

summer, and for the south wall during the winter. Good building planning can take this into account by minimizing the east-west exposures and using the south exposure for heat gain during the winter. Exterior shading devices and trees can also be used effectively to control solar radiation.

Heat gains and losses from conductive transfer through walls exposed to outside air are dependent upon the resistance of the wall materials to heat transfer and upon the temperature differential between the inside and outside air. The transfer across these materials is measured by the U factor of the composite wall construction. A minimum U factor will minimize these types of gains and losses. Generally, an air space is included in wall construction because still air is an excellent insulator and substantially lowers the U factor.

Heat gains and losses from infiltration of outside air occur through unplanned and uncontrolled openings in the building envelope. This accidental airflow through the building envelope carries with it considerable heat. It can have a substantial effect when the inside-outside temperature differential is great. Infiltration is controlled by good joint design and adequate weather stripping.

Heat gains and losses from ventilation with outside air result because there is always a need to bring fresh air into a building. The occupants and internal machinery produce a considerable need for regenerating the interior air. Additionally, heat gains from internal sources such as computers have become quite significant. These internal gains are in part controlled by effective ventilation but are also controlled by reducing operating levels to a minimum.

A building can be thought of as a system of interacting energy flows. The system can be seen graphically to be an interaction of the five transfer mechanisms discussed above. The designer-contractor seeking to make buildings as energy-efficient as possible needs to consider fully the implications of Fig. 1-2.

1-5 THERMAL INERTIA OF CONSTRUCTION MATERIALS

The traditional masonry wall provides us with an excellent example of a heat storage reservoir. Its so-called thermal inertia results from its inherent high density (or mass), which responds favorably to constant variations in the ambient air temperature as it fluctuates during each 24-hour cycle. This cyclical effect is graphically illustrated in Fig. 1-3. Note that the ambient temperature generally rises to some maximum temperature during afternoon hours and falls to some minimum at night. The space air temperature in an unheated or unconditioned building will follow a similar cyclic trend, with the range of temperature extremes somewhat smaller due to the time lag of the materials in reaching these ambient extremes.[4]

The greater the mass of walls and roof, the longer it takes to heat or cool the exterior building envelope. As a result, a light-weight structure will exhibit faster and greater changes in indoor temperature than heavier structures, since it can be heated

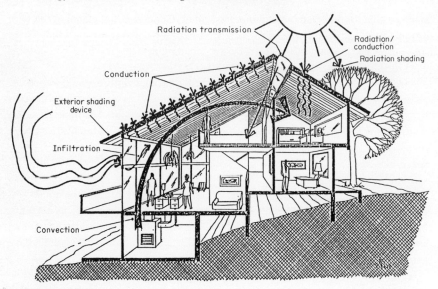

Fig. 1-2 Heat transfer in buildings. (The Energy Group, Consultants, Los Angeles, California.)

up or cooled down faster by the heating, ventilating, and air conditioning (HVAC) systems. The resistance of an exterior building element to outside cyclic temperature changes has been defined as the *thermal inertia* of a wall. In concept it is somewhat different than a wall's conductance or U value. Conductance, expressed in Btu/(h)(ft²)(°F)[W/(m²)(K)], can be used to determine the rate of heat flow through a material when the temperature difference on each side remains constant (a steady-state condition). The property of thermal inertia can be used to determine how fast the wall will heat up or cool down (its dynamic response). This is dependent upon wall thickness, density, specific heat, and conductivity. Since the ambient is constantly changing, thermal inertia is an important consideration in studies of heating and cooling.

The cost of heating and cooling within various building types is materially affected

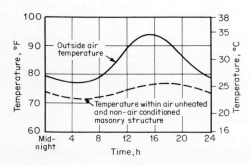

Fig. 1-3 The outdoor temperature is constantly changing. ("Energy Conscious Design for Buildings," NCMA-TEK 82, National Concrete Masonry Assoc, 1976.)

by thermal inertia. An increase in the thermal inertia of a wall results in a decrease in energy cost. In contrast to this, a reduction in the U value of the wall may, in some cases, increase the total annual cost of heating and cooling.

When the thermal inertia of the wall is taken into account, the calculation is said to be dynamic. When thermal inertia effects are ignored, the calculation remains a static approximation.

1-6 COOLING CRITERIA AND ENVELOPE MATERIALS

The proper design of modern building air conditioning systems demands a broad appreciation of transient heat flow phenomena in construction materials. Differences between inside and outside temperatures and solar radiation rates are highly variable during the course of the day and, consequently, set up periodic heat flows through the exterior construction.

Actual instantaneous heat gains do not reflect immediately on HVAC equipment. They must first be absorbed as radiant or convected heat. This ultimately finds its way by conduction to interior surfaces to be convected to the circulating space air. A lag results between the time that the incident load is received by the exterior building construction and the time when all or some portion of it is transmitted to the conditioned space.

The thermal storage characteristic of the exterior building envelope is of great interest in building design.[5,6] While its value varies periodically during the course of a day, primarily as a result of outdoor conditions, it is also modified to some extent by the programmed building occupancy with regard to lighting and equipment use.

1-7 BUILDING ENVELOPE SOLAR ENERGY TRANSFER

During the fifties and most of the sixties, architects concentrated on mastering nature, independent of favorable (or unfavorable) microclimate conditions. It was unthinkable in terms of availability or cost to question the use of often unreasonable quantities of energy needed to maintain close control of interior environments of buildings designed without regard to thermal efficiency or reasonable fenestration. Furthermore, overdependency upon building air conditioning systems in unshaded, essentially sealed-glass structures which become uninhabitable shortly after equipment breakdowns, has often strained owner-tenant relations. Excessive air motion and window wall draperies which conceal desirable views for extended daily periods often degrade the quality of an otherwise cherished executive suite. When draperies are occasionally parted and sun rays strike the occupant, the thermostat, sensing only the prevailing space air temperature, cannot restore comfort. Perhaps we should be

12   Energy Conservation in Buildings and Industrial Plants

grateful to the 1973 oil embargo for bringing us to our senses and forcing us to design our buildings with nature (see Fig. 1-4) instead of consigning vast amounts of energy to conquer it.

It is helpful to consider our building as a *free body* so that the dominant energy flows can be traced to early design decisions. For example, let us compare two concepts for level high-rise office building structures[1] of the type illustrated in Fig. 1-5. Building A represents a conventional approach with curtain walls comprising unitized insulated metal panels and glass sections uniformally covering the entire

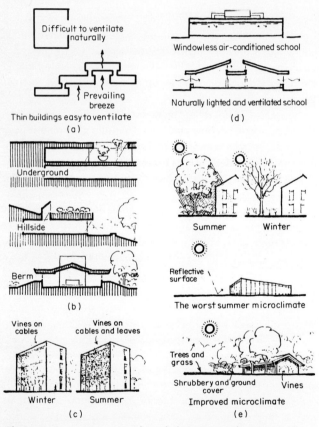

Fig. 1-4 Contrasting examples of designing with nature: (a) arrangement to maximize ventilation flow; (b) planting the structure in the ground to conserve energy; (c) use of decorative plants on buildings; (d) mechanical versus natural ventilation potentials; (e) improvement of the microclimate. (C. W. Brubaker, "Energy Conservation through Rational Architecture and Planning," *Building System Design*, June/July 1975.)

Conservation in Exterior Design and Construction   13

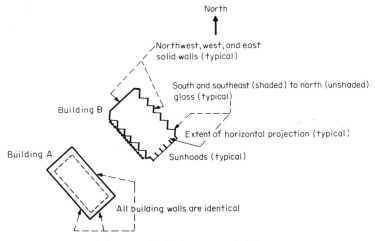

Fig. 1-5 Analysis of alternate (freebody) concept buildings.

exterior. Building *B*, however, represents an energy conservation approach to the same design problem. It is a composite structure articulated with solid walls and glass sections arranged to maximize the view and minimize adverse impacts of the sun.

As the morning sun rises in the east (higher in the summer than in winter months) on building *A*, solar radiation strikes the east curtain wall. Interior draperies or shades must be drawn to avoid the low-angle sun rays. During the winter months the draperies tend to reflect and localize solar heat gains otherwise transmitted through the glass. The insulated thin-wall (low-mass) panels tend to reduce the inward conduction of solar radiation striking the outer walls and to minimize retention of this heat.

As the sun continues to move through the south to the west, the south face of building *A* will experience a maximum solar radiation flux near noon. As afternoon hours arrive, the sun falls in the west and higher afternoon temperatures (particularly during the summer months) build up heat gains during the late afternoon hours. Building *A* represents a product of earlier times when perhaps outward appearance dictated a uniformity in style.

Now examine building *B*, which represents an attempt to relate to natural daily sun patterns. It provides a more energy-efficient solution (hopefully without sacrifice of aesthetic qualities) by recognizing the need for different wall treatments depending upon exposure. Notice that the early morning sun warms the dense solid (east) walls which are designed to retain much of the absorbed (solar) heat.

Noon sun is controlled on the southeast and south by exterior horizontal-vertical shading devices. The afternoon sun, which can be bothersome, impinges upon dense solid (west) walls which also retain this absorbed solar heat gain. Notice that the southwest glass is shaded by both horizontal fin elements. Only the north-facing glass, which provides pleasant year-round views, remains unshaded.

1-8 SUN HOODS: A VIABLE PASSIVE ALTERNATIVE

The use of structural glass on exterior walls has become a widely accepted practice, particularly in high-rise construction. Despite the availability of structural glass in a variety of types, colors, and shapes, its indiscriminate use often necessitates the installation of an elaborate mechanical system. This may cause the building owner a higher than necessary first cost or increased operating costs. When large quantities of structural glass were introduced some years ago, many of the mechanical building systems then available proved inadequate.[7]

Improved technology has eliminated many of the comfort and control problems resulting from the earlier use of non-load-selective-type[8] HVAC systems serving essentially floor-to-ceiling glassed areas. Nevertheless, it is still wasteful to permit excessive amounts of solar energy to enter a building envelope, regardless of how efficiently such loads may be removed or redistributed to offset winter heating[9] demands in the northern climates or used for daylighting.[10]

Conventional exterior sun-shading devices are distasteful to many architects, who look upon them as unsightly appendages. An alternative approach is to employ sun hoods that derive their shape from the natural requirements of the building, its geographic location, and its orientation on the site. Furthermore, such sun hoods can be detailed as an integral facade design element that can often enhance the appearance of the building.

The concept of the sun hood is best described by analogy[9] to a well-shaped shade tree covering an area of structural glass as the sun moves across the sky. Long before sophisticated buildings filled our horizons, a good shade deciduous tree was a coveted asset (see Fig. 1-4). It served to intercept direct incident solar radiation at the time of day when the sun's rays were most intense. Yet it permitted some solar radiation to pass through when supplementary heating was advantageous. This principle can be readily applied to the contemporary design of a building facade by introducing projections that provide a shade profile to compensate for the increasing intensity of incident solar radiation as the sun moves across the sky on the design cooling day. In effect, sun hoods act as time barriers that intercept solar radiation in direct proportion to its intensity. Sun hoods thereby reduce the solar flux (energy) that can enter a building during hours that correspond to its maximum cooling load. It is this short span of time (referred to as the *maximum coincident cooling-load period*) that establishes the required capacity of the building's mechanical cooling systems. Solar peaks can be trimmed through careful study of each individual building-to-sun relationship.

1-9 COMBINATION PASSIVE-ACTIVE CONSTRUCTION ELEMENTS

Another technique, still in the experimental stage, which holds promise for the regulation of solar flux into buildings requires the use of thermal panels[12] in which a given wall or roof envelope element becomes a thermic diode. As such, the wall or roof element performs the following functions: (1) collection of (external) solar or dissipation of (internal) heat gains, (2) retention of heat within the panel, and/or (3) control of the direction of heat flow to and from the panel.

The thermic diode approach[13,14] is somewhat different from other solar (heating) systems, some of which will be described in more detail in Chap. 7. For the purposes of comparison, it may be helpful to make some general distinctions. Active solar systems, either of the air or water type, require fans and/or pumps for the distribution and heat transfer of collected heat gains. Passive systems, much like the wall envelope masses described earlier, employ only the sun's energy for collecting and storing solar heat, and are often built into the structure itself. Thermic panels do not fit well into these categories, yet they have elements of both. Thermic panels employ a low-volume water storage medium as do both active and passive water heating systems. Yet, not unlike passive systems, thermic panels have an integral storage medium which reduces material and installation costs. Compared to the lowest first-cost active system, such as air-type solar heating systems, thermic panels installations can be expected to cost only two-thirds as much. When compared to water-type (heating) active systems, a reduction of 50 percent can be anticipated. Figure 1-6 illustrates some approaches to incorporating thermic panels as part of a conceptual design strategy. During winter, heat is provided by convection (between

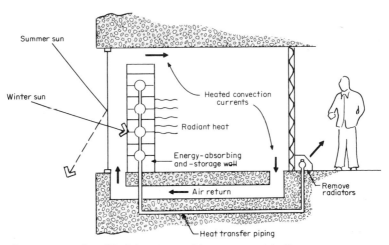

Fig. 1-6 Use of modified thermic panel in a commercial office structure.

the glass and the thermic panel) and by direct radiation from the panel into the room. Additionally, the hot water system built into the panel provides heat for remote radiators. In the summer, curtains are drawn behind the window to render the system inactive.

A thermic panel of the type illustrated in Fig. 1-7 is designed to accommodate sufficient water in its internal passages to store approximately one day's heat requirements. Such a thermic diode controlled within the panel is schematically illustrated in Fig. 1-8. Four modes of operation are possible, with heat flowing in only one direction at a time, but capable of automatic switching to a preferential direction of heat flow depending on prevailing outdoor temperatures. The thermic diode control box consists of an oil check valve which responds to very slight temperature-induced pressure conditions. The valve seals completely in the reverse direction. Notice, for example, that during the winter the panel can absorb and store solar heat gains internally while permitting physical isolation of the stored heated water from contact with colder exterior surfaces when there is no sun. In the summer, these same panels serve to provide insulation from the sun's rays while cooling interior areas with stored water cooled by transfers during the preceding night.

While thermic panels remain somewhat experimental, they illustrate the need to integrate passive systems and dynamic systems, such as heating and air conditioning, into an overall design approach. Passive systems and dynamic systems should complement each other as much as possible.

While devices such as sun hoods and, possibly, thermic panels may be very effective in appropriate situations, it must be recognized that more fundamental changes in building design will be needed if we are to utilize our energy resources more efficiently in the future.

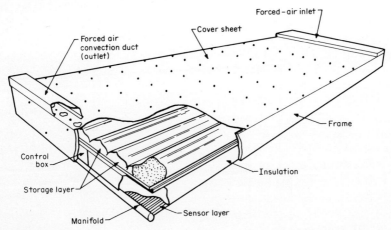

Fig. 1-7 Prototype test panel. (Shawn Buckley, "Thermic Controls to Regulate Solar Heat Flux into Buildings," National Technical Information Service, U.S. Department of Commerce, August 25, 1975.)

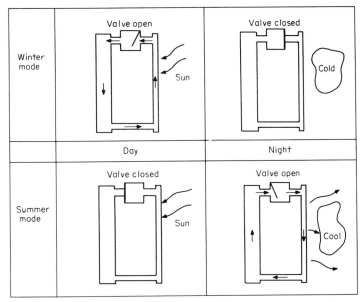

Fig. 1-8 Simplified schematic of heating and cooling modes. (Shawn Buckley, "Thermic Controls to Regulate Solar Heat Flux into Buildings," National Technical Information Service, U.S. Department of Commerce, August 25, 1975.)

1-10 ENERGY-EFFICIENT STRUCTURES

An exciting area of new development in structures involves the so-called *self-built* structures anticipated near the turn of the century. Such structures differ substantially from those involving the contemporary passive materials discussed earlier. Instead, they deal with active materials into which energy may be constructively introduced to impart exceptional strength and stiffness.[1,15]

In nature, most biological structures appear naturally thick over highly stressed segments and thin at lightly stressed portions. We are now trying to understand how such natural processes work so that some duplication can be incorporated into artificially made structures.

Pneumatic or hydraulic structures are one of the most fundamental forms found in nature, with wide application in plant and animal life. Such structures are commonplace in fruit, air bubbles, blood vessels, and in skin kept taut by muscle tissue and blood pressure. Yet, although the mechanical utility of hydraulic and pneumatic systems has been apparent for some time, only recently has serious consideration been given to utilizing hydraulic[16,17] and pneumatic concepts[18-21] in actual building structures. Although the more traditional architects and structural engineers view the use of air-supported structures with some skepticism, the increasing use of so-called *bubble buildings* persists for a variety of low-rise structures, including convention and sporting arenas and temporary and exploratory shelters.

Similar adaptations may use hydraulic structures employing liquids since (1) liquids (and gases) are able to transmit force levels instantaneously, equally in all directions, and without stress concentration in the building envelope for adequately designed structures; (2) networks encasing the hydraulic fluids experience tensile forces under load, and normally impinged compressive loads are transformed to fluid pressures which are resisted by the network, e.g., container; (3) load-bearing capacities can be distributed throughout the network by controlling relative area(s) on which such loads are reacted by the hydraulic network.[22,23]

Pneumatic or hydraulic structures comprise structural frameworks which are stabilized, reinforced, or supported to some degree by pressure differences. In all instances, we find the essential features of a light-weight structure with (1) tension forces (stress) reacted by pneumatically slightly prestressed skins and fibers which enable them to support broadly distributed surface loads and (2) compressive and bending (forces) which are concentrated locally.

Tubular members filled, under pressure, with liquid can resist considerably higher forces than conventional solid members.[16,17] This increased capacity is due to the prestressing effect of the pressurized tubular members, and also to the increased resistance to buckling of the columns and other compression members.

Hydraulically pressurized structural members may be an improved means for designing structures that economically resist wind and earthquake forces. On structures in excess of 40 stories, for example, wind forces often approach seismic forces in magnitude.[17]

Let us briefly examine how such a system might work. In Fig. 1-9 on page 20, we have a tubular element, uniform in cross section, in which a prestress has been applied by internal fluid pressure to a level just sufficient to assure that the net axial stress in the tube wall will not become compressive upon the application of an axial load A. Assume further that the tube element is just one of many making up a typical column, thus able to transmit load A across a column joint. The infinitely thin membrane shown at either end merely defines a boundary condition necessary for purposes of analysis.

In columns constructed of a plurality of fluid-filled elements, the total combined weight is not only a function of the tube material, the fluid medium, and length but also of shear stress and the internal fluid pressure. When we compare a long pressurized tube with a conventional short, solid structural column of the type normally found in multirise buildings, we see that pressurization leads to roughly one-third savings in the weight of the shell and a potential cost savings therefore of one-third.[15] Furthermore, when using a glycol-water solution as the pressurization medium, we gain the further advantage of fire protection. A weight savings should always result[15] provided:

Conservation in Exterior Design and Construction 19

1. The ratio of the pressure level to the fluid density is always greater than the ratio of uniaxial yield strength to the material density of the solid column.

2. The tensile strength of the material is approximately two times its compressive strength (or greater).

In short, prestressed columns favor low values of structural index. Thus, longer column lengths become more economical than solid column members, which favor high structural indexes. In the pressurized tube, the potential weight savings can be appreciable. The weight of the pressurized column varies directly with the applied load and the column length but is surprisingly independent of the modulus of elasticity. For this reason, almost any gastight material can be used, although materials of high modulus of elasticity may provide more compact members.

The idea of using fluid in structural members has already been tried in several buildings[24] in the United States, the most dramatic being the recent 850-ft-high (259 m) U.S. Steel Headquarters building located in Pittsburgh. The buildings' 18-box columns, 2 by 3 ft (0.61 by 0.91 m) in plan sections, are filled with fluid primarily for fire protection. Each column is divided into four vertical zones by solid diaphragms, and each zone contains about 500,000 gal (1,892,706 L) of water and 740 tons (671 t)* of potassium carbonate to prevent freezing.

The concept of the energized column has already been advanced for space structures,[15] where the circulation of fluid within a column is capable of supplying a portion of the column's strength. With an ideal fluid, recirculation equally top to bottom might well enable even greater reductions in total column weight. It has been suggested[15] that, if the means were provided to recirculate an ideal fluid within a structural column equally top to bottom, the weight of the fluid and the column combined would approach levels of 25 to 30 percent of the prestressed column illustrated in Fig. 1-9.

A conventional steel column and beam framing system utilizing solid steel members results in a significant loss of available material strength when compression forces are considered. The result often is a structure which occupies more volume and is heavier than necessary; most of the compressive force could be transferred into the wall of an equivalent steel tubular column with characteristically higher volume-to-weight and volume-to-stress values.

Other wasteful practices relating to stress problems abound in the design of multirise buildings. In the proportioning of columns, for example, comprehensive loads must be transmitted over large distances. Traditional methods do not fully utilize the inherent strength of materials employed; to provide stability, structural engineers are required by code to oversize columns by using stress values which are low compared to ultimate strength.

In the years to come, we will need to rethink the entire way in which we use basic

---

* The abbreviation "t" is used to designate metric tons throughout this book.

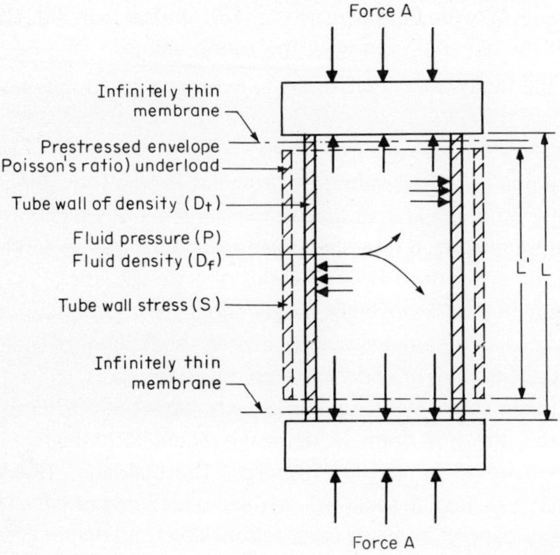

Force A

Infinitely thin membrane
Prestressed envelope (Poisson's ratio) underload
Tube wall of density (D_t)
Fluid pressure (P)
Fluid density (D_f)
Tube wall stress (S)
Infinitely thin membrane

L' L

Force A

Total weight of element (tube + fluid) = $\dfrac{ALD_t}{S}\left(2+\dfrac{SD_f}{PD_t}\right)$

Fig. 1-9 Analysis of prestressed tubular element. (Milton Meckler, "Fluid Filled Tubular Tension Structures," Civil and Structural Engineering Proceedings, *Dialogue in Development: Natural and Human Resources*, Third World Congress of Engineers and Architects, Tel Aviv, Israel, December 1973.)

raw materials. Once again, for reasons of economics and because of problems with supply, new, more flexible attitudes will be required.

SUMMARY

Because of the increasing shortage of inexpensive energy sources, energy conservation is no longer optional. To understand how to implement such conservation measures, the energy planner must understand the basic principles of heat transfer: heat conduction, heat convection, and heat radiation. The energy planner must also be familiar with the concepts of thermal mass and thermal inertia.

In addition to the thermal quality of building materials, building orientation with respect to the sun must be considered. Passive devices such as sun hoods and promising new passive-active options such as thermic panels should be investigated. The prudent energy planner should strive to optimize the balance between passive options and dynamic options.

In the future, fundamental changes in building design will be needed if we are to utilize our energy resources effectively. Pneumatic or hydraulic structures appear promising.

During the coming years, planners and designers must rethink the entire way in which we use basic building materials. Forecasts indicate a probable lumber shortage during the next decade, and structural plastics have failed to live up to their earlier promise. Steel and concrete will remain in sufficient supply through the year 2000, and significant improvements in their structural qualities are anticipated. Increasingly, the amount of energy needed to fabricate the various construction materials will become an important factor in building design. Likewise, the amount of energy expended on the building site during construction will take on added significance.

NOTES

1. "Probing the Future," *Eng. News-Rec. Centennial Issue*, **192**(18) (1974).
2. Robert A. Kegel, "The Energy Intensity of Building Materials," *Heat./Piping/Air Cond.*, pp. 37–41 (June 1975).
3. "Energy Use in the Contract Construction Industry," U.S. Department of Commerce Report no. PB-245-422, Feb. 18, 1975.
4. Francisco N. Armui, "Thermal Inertia in Architectural Walls," paper presented at the Annual National Concrete Masonry Association Federal Offices Lectures, July 1976.
5. M. J. Catani and S. E. Goodwin, "Thermal Inertia—The Neglected Concept," *Constr. Specifier* (May 1977).
6. Mario Catani and Stanley E. Goodwin, "Heavy Building Envelopes and Dynamic Thermal Response," *J. Am. Concr. Inst.*, p. 83 (February 1976).
7. M. Meckler, "New Guidelines for the Owner Who Builds," *Buildings* (April 1965).
8. M. Meckler, "Load Selective Multizone Systems Condition Growing Telephone Building," *Air Cond., Heat., Vent.* (September 1964).
9. B. L. Collins, "Evaluation of Human Response to Building Fenestration," Paper HA-77-4, no. 1, National Technical Information Service.
10. J. W. Griffith, "Benefits of Daylighting—Cost and Energy Saving," Paper HA-77-4, no. 2, Halifax ASHRAE, National Technical Information Service.
11. "Concrete Trees for Shade," *Interbuild, Prefabrication Publications Ltd.*, p. 38, London (July 1965).
12. "Thermic Controls to Regulate Solar Flux into Buildings," National Science Foundation Final Report, April 1976, NTIS PB-2533345, National Technical Information Service.
13. W. Stargardt and S. Buckley, "An Economic Analysis of Thermic Diode Solar Panels," presented at American Society of Mechanical Engineers, Winter Annual Meeting, New York, December 1976, 76-WA/SOL-7.
14. "Thermic Diode Solar Panels, A Brief Summary," *Proc. Int. Solar Energy Soc. Joint Conf.*, Winnipeg, Canada, **2**:1–23 (August 1976).
15. Ralph L. Barnett, "Optimum Prestressed Tubular Columns," *J. Struct. Div. Am. Soc. Civ. Eng.*, **96**(ST2) (February 1970).

16. M. Meckler, "Pressurized Tubes Within Columns Support Multi-Story Buildings," *Hydraul. Pneum.* (April 1974).

17. M. Meckler, "Fluid Filled Tubular Tension Structures," *Civil and Structural Engineering Proceedings, Dialogue in Development: National and Human Resources,* Third World Congress of Engineers and Architects, Tel Aviv, Israel, December 1973.

18. Jens G. Pohl, "The Structural Prestressed Flexible Membrane Column," *Archit. Sci. Rev.* (June 1970).

19. Jens G. Pohl, "Pneumatic Structures," *Archit. Aust.,* **57**(4):635–639 (August 1968).

20. Jens G. Pohl, "Air & Water Support Structures," *Build. Mater. Equip.,* pp. 37–41 (June/July 1972).

21. Otto Frei, *Tensile Structures,* M.I.T., Cambridge, Mass., 1967.

22. U.S. Patent No. 3,538,653.

23. U.S. Patent No. 3,796,017.

24. "A Trio of Unusual Structural Concepts," *Build. Des. Constr.,* p. 47 (February 1974).

Energy Conservation Strategies in Interior Design

The one factor that, more than any other, determines energy consumption of a building is how it is used.

Lawrence G. Spielvogel, P.E.

Recent opinion polls indicate a majority of consumers cite higher costs as the basic rationale for conserving energy. Economics, not education, is believed to be the basic impetus.[1] Therefore, to be creditable, an energy conservation study (ECS) must be carefully aimed at reducing energy costs on a priority basis against additional capital expenditures which may be required.

2-1 CONSIDERING AN ENERGY CONSERVATION STUDY (ECS)

In determining whether a building is a candidate for an ECS, we may find the following checklist helpful:

1. How does the existing building compare to similar buildings in its area? Does it consume more metered energy, and, if so, are there any unusual operating or construction factors which might help explain it?

2. If item 1 raises serious doubts, then we must proceed to accurately "model" (in computer jargon) the building and installed systems to determine theoretical energy consumption levels. This establishes a general level of expectations.

3. Once we have completed item 2, we can consider modifications to operating and design parameters. By studying theoretical system performance

to provide meaningful data for our "actual building" and by various calculations, we can begin to simulate the probable effect and energy-saving consequences of numerous alternates.

2-2 THE ANNUAL ENERGY BUDGET

It may be helpful at this point to introduce the concept of an annual energy budget (AEB). The AEB is defined as the amount of fuel at its gross thermal value used by a building, added to the kilowatthours of electricity used, times a factor of 3143 Btu/kWh (33.16 kJ/kWh). This total is then divided by the square footage of the building.

AEB information is generally available for most buildings since information normally supplied with utility bills permits direct computation of this value. It should be pointed out, however, that seldom are individual items of equipment monitored, and the quality of information, particularly in older buildings, is often questionable. Information involving older building use is often difficult and expensive to obtain. Therefore, reported data must be carefully examined as subsequently described, taking into account their source, the means of data collection, and the validity of the sample.

Significant variation in estimated AEBs often results when a building HVAC system design or operation is altered. In some cases, variations of from 5 to 10:1 have been found, depending upon the range of HVAC systems considered and the proposed operational features.

Let us examine some revealing general statistics regarding office buildings provided by the Building Owners' and Managers' Association (BOMA). Available AEB data, segregated for the various general building categories and even climatic factors, disclose a rather broad range in AEBs.

Refer to Fig. 2-1 for a plot of air conditioning costs versus degree-days for 125 million ft² (11.61 million m²) of office space. Note that there is little if any correlation between climate and air conditioning operating costs without some corrections for the influence of operating hours.

Another interesting statistic is evident after a cursory inspection of Fig. 2-2; there appears to be no direct correlation between annual energy use for lighting and the watts per square foot (watts per square meter) of installed capacity.

What about using the efficiency of HVAC operating equipment rated at full load as an indicator of how well the building can be operated? The normal centrifugal chiller may require 0.7 kW/ton (0.2 W/W) while some less efficient units, such as package heat pumps and air conditioners employing small hermetic compressors, operate at 2 kW/ton (0.57 W/W). At first glance, this might indicate an overwhelming advantage to the centrifugal chiller. But is this so? Experience often indicates otherwise. Consider the following circumstances: We have a hypothetical 450,000-ft² (41,805-m²) office building with two 750-ton (2370-kW) centrifugal chillers. Assume that any one tenant wishes to work late some evening and that, in addition to the one

Energy Conservation Strategies in Interior Design 25

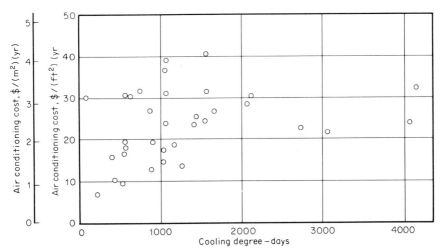

Fig. 2-1 Plot of air-conditioning cost versus cooling degree-days for 125 million square feet of office space. (Lawrence G. Spielvogel, *Exploding Some Myths about Building Energy Use.* Reprinted from *Architectural Record*, February 1976, by McGraw-Hill, Inc.; all rights reserved.)

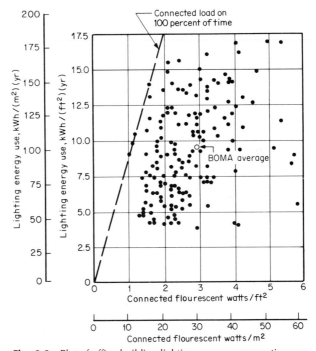

Fig. 2-2 Plot of office building lighting energy consumption versus installed capacities. (Lawrence G. Spielvogel, *Exploring Some Myths about Building Energy Use.* Reprinted from *Architectural Record*, February 1976, by McGraw-Hill, Inc.; all rights reserved.)

TABLE 2-1 Energy Usage in a Shopping Mall

	Btu/(ft²)(yr)	kJ/(m²)(yr)
Auto center	74,000	839,821
Department store	114,000	1,293,778
Department store	102,000	1,157,591
Variety store	100,000	1,134,893
Restaurant	409,000	4,641,712
Bank	131,000	1,486,710
Drugstore	129,000	1,469,012
Food market	205,000	2,326,530
Dry cleaner	688,000	7,808,063
Book store	104,000	1,180,289
Doughnut store	326,000	3,699,751

SOURCE: Lawrence G. Spielvogel, "Exploding Some Myths about Building Energy Use," *Architectural Record* (February 1976). Copyright © 1976 by McGraw-Hill, Inc. All rights reserved.

chiller, a couple of 150-hp circulating water pumps, a 75-hp cooling tower fan, one 500-hp primary-air fan, and one 150-hp return-air fan are all required to run simultaneously. All told, approximately 1500 kW is being consumed to provide conditioning for a common, after-hours work situation normally requiring about 2 kW per occupant. This problem could be minimized, on the HVAC system air side at least, by better zoning and modular central plant design. While we are not attempting here to make a case for unitary versus the more efficient central plant systems, the above example does point up the need to recognize early in HVAC design and system planning that building use is a primary determinant of energy use.

Stores and other retail establishments are often difficult to compare on the basis of AEBs. There are wide variations depending upon use even within a given microclimate as illustrated in Table 2-1, taken from an enclosed-mall-type shopping center located 50 mi (80 km) north of Philadelphia. Each store is served from one or more electric heating and cooling unitary packages. Stores are individually metered and directly billed. All stores operate a fixed number of hours per day and days per week. In spite of this, we find a 10:1 ratio in AEBs. It is reasonable to assume, therefore, that store functions play a key role in energy use.

2-3 THE ENERGY FLOW SHEET

Unfortunately, there is little detailed data on energy flow within various types of buildings and their operating systems. Reliance on metered utility billings at the building project boundary can be misleading, since it reflects a total rather than a breakdown of the specific energy used by each building system. Often it becomes difficult to know whether the energy consumed is materially affected by the installed HVAC system or by some anomaly.

One suggested approach[2] is to develop a graphic representation of building energy flows on an energy flow sheet by accounting for all the energy which flows through a building and is converted and finally expelled to the earth, atmosphere, or water surrounding the building. The energy flow sheet is an attempt to take into account the complex interactions which define the utilization of energy as it flows through the various major building systems or components. In Fig. 2-3, notice the high-grade energy sources, such as electricity and No. 6 oil or gas, entering the building project boundary (depicted by a dashed line). There are incidental energy inputs shown which result from ventilation air, water makeup, solar gains, and the building occupants. Also shown is the low-grade energy lost to the atmosphere as described earlier.

The left-hand portion of Fig. 2-3 represents energy flows to the central plant or generation uses, while the right-hand portion deals with various energy and uses or loads. The various nodes represent items of equipment, or portions of the building, and the energy flows which relate to them. The interconnecting lines between nodes indicate the flow of energy in millions of Btu per hour (megawatts) from one node to another. The energy flow sheet can be thought of as a map of energy use within a building.

The flow sheet in Fig. 2-3 is representative of a typical summer design day averaged for a 24-hour operating period.

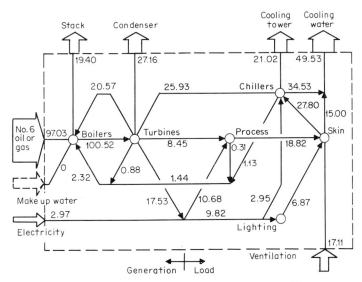

Fig. 2-3 Sample energy flow sheet. (James L. Coggins, "An Effective Energy Audit: The Flow Sheet Approach," *ASHRAE Journal*, June 1976.)

It may also be useful to develop energy flow sheets for other building occupancy time periods. For example, during minimum energy demand periods, energy flow sheets can be used to reconcile or evaluate various distribution system thermal losses. The enthalpy or heat content of the makeup water to the boiler can be used as the energy base to simplify computations.

To obtain values for constructing energy flow sheet, we must often draw on several information sources. Utility bills, submetering and other operating records, central power plant and equipment logs, charts, contract documents, balancing reports, and independent surveys may all prove helpful. The importance of the independent survey cannot be overestimated. In such a survey, key items such as temperatures, lighting levels, electrical consumption and demand, and boiler and chiller efficiencies must be measured. After the necessary data have been assembled, we enter this type of information on the energy flow sheet. Then, starting at any given node, we will usually find that most but not all of the energy flows are known. However, by repeating the process, we can usually determine the unknown values necessary to make each of these nodes balance. Furthermore, the energy flow sheet provides a frame of reference from which most differences between information gathered by us or supplied by others are reconcilable.

If we cannot achieve a balance, it may be necessary to return to the data source and review key assumptions. By assembling the data necessary to construct the energy flow sheet, we are automatically forced to:

1. Correlate and reconcile sketchy or patchy data.
2. Determine where energy is used and establish the major energy flows or losses.
3. Determine the energy leverage factors.
4. Organize the available data and point out any gaps in additional required information.
5. Provide a cross-check of available information to improve its reliability.
6. Make evaluations among various proposed energy conservation measures (ECMs).

Admittedly, the energy flow sheet tends to oversimplify some problems, but it can be useful in structuring an energy audit and as a rational framework for comparing various alternative ECMs.

In a subsequent chapter, we will explore a further use of energy flow sheets in comparing energy flows beyond the building project boundary to develop improvements on ultimate fuel resources.

2-4 EXAMINING SOME MYTHS ON BUILDING ENERGY USE

While it might seem that equipment design factors in HVAC systems, boilers, etc., should be more important than the uses to which the equipment is put, the reverse is actually true. Regardless of the fenestration, insulation, or lighting within a building, it is the number of operating hours of HVAC system use that most often determines a building's energy consumption. Therefore, merely comparing buildings on the basis of installed equipment or lighting capacities or design heating and cooling loads can prove misleading.

Another common myth concerns the effects of weather. Only a small fraction of the annual energy use occurs during the extremes of weather. Most of the energy consumption in buildings occurs during moderate ambient conditions, corresponding to the operation of building HVAC systems at part-load most of the time. Thus, we must become more concerned about HVAC system efficiencies and building envelope factors during moderate weather.

Another common source of problems associated with energy use involves a building's spatial organization and the different functions and periods of use. One must differentiate spaces used only occasionally, such as auditoriums and conference rooms, from those, such as computer rooms, requiring 24-hour use. A recent study of 50 office buildings in Philadelphia[3] revealed that "the variables most affecting energy consumption are the extent and type of building use, as determined by presence or absence of computers, data processing equipment and support facilities." Energy use was found to be 50 percent higher in comparable buildings with such extended-use facilities. Furthermore, when operating 168 hours a week, a building required additional lighting in lobbies, stairwells, corridors, lounges, and elevators for personnel. Such building services for maintaining the so-called building "spine" often approach the energy requirements of fully occupied buildings. In fact, computer rooms occupying less than 10 percent of the available floor area have been reported in some cases to have consumed more than two-thirds of the annual electrical usage. Therefore, it is necessary to carefully examine the proposed building usage and arrange HVAC, lighting, and other building services in such a way as to avoid the wasteful operation of building systems. Such situations are easily remedied during the initial design stages by more flexible zoning, diversity, etc.

2-5 IMPACT OF THE INTERIOR DESIGNER

A third of the nation's fuel is required annually to operate residential and commercial buildings.[4] This fuel is wastefully consumed, partly because conservation has not been a consideration in the design of built environments until recently. According to the current studies of GSA, FEA, and NBS,[5-7] the greatest energy conservation opportunities in commercial facilities are derived from astute managerial practices applied to an entire facility and to the interior environment in particular. Savings are greatest when an interior environment has been designed with an emphasis on energy conservation. Although buildings that require an absolute minimum of fuel energy are necessary,[7] little has been done to draw energy conservation design techniques into the working practices of the professions responsible for the design of the built environment.

It is here that the interior planning and design profession can play a pivotal role. The planning–programming–design–industrial-engineering process provides most of the necessary information for the design of facilities in accordance with the occupants' work and personnel flow requirements. The interior planning and design profession creates physical forms to accommodate the functions of a commercial concern. To quote the maxim, form follows functions. The planner-designer synthesizes the output of all related professionals who contribute to the design of the interior environment. The planner-designer sets the parameters within which related professions may base their individual design criteria. The planner-designer is, in turn, responsible to the architects who produce finished building shells.

See Fig. 2-4 for a comparison of a so called "standard" approach to building design and a proposed design approach for maximizing energy efficiency. Substantial increases can be achieved by relating building shells to a prearchitectural assessment of user requirements and options. Although conservation opportunities are greatly diminished by the limitations of a fixed building shell, energy consumption can still be reduced approximately 15 percent by a mix of interior technical

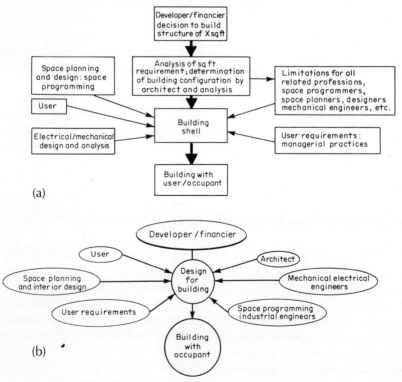

Fig. 2-4 (a) Standard approach to building design. (b) Design approach for maximum efficiency. (Milton Swimmer Associates, Beverly Hills, California.)

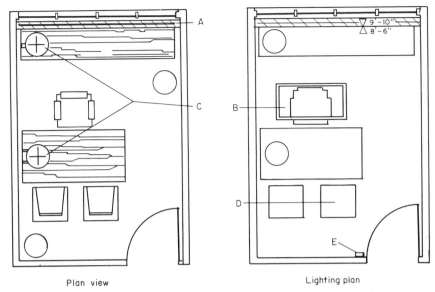

Plan view   Lighting plan

Fig. 2-5 Typical 14′ × 10′ (4 m × 3 m) office-collective design options. (A) Vertical louver window treatment. Louvers are dark on one side to draw in winter sun, light on the second side to reflect summer sun. Vertical Louvers are commercially available; dark- and light-sided louvers are a project innovation. (B) Air-handling fluorescent light fixture to use heat from light fixture for HVAC system. (C) Task lighting to illuminate the direct area in which work is done. (D) Air-handling incandescent light fixture. (E) Selective switching for light fixtures to allow option to choose specific light fixtures and brightness of light for each task. (Milton Swimmer Associates, Beverly Hills, California.)

options. The use of interior design standards of the type illustrated in Fig. 2-4 can ensure that conservation opportunities are considered when space layouts are prepared. A concerted effort to introduce pragmatic exterior and interior planning will have a great cumulative affect on the built environment.

Figures 2-5 and 2-6 illustrate a collection of technical options that may be used individually or in unison to conserve energy in a typical office.

2-6 THE ROLE OF THE ELECTRICAL DESIGNER

The development of effective programs to conserve energy in new and existing buildings requires that the electrical designer initiate innovative approaches in the design of illumination and power systems.

Automatic control of lighting is recommended for well-fenestrated buildings. A simple study is usually recommended first, to determine "on/off" available light levels for a given building or portions thereof.[8]

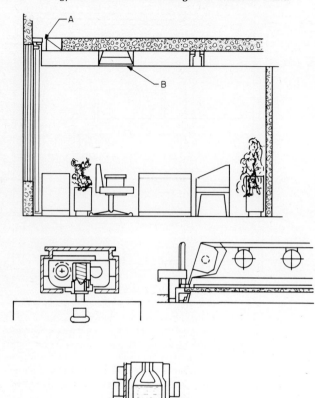

Fig. 2-6 Typical 14' × 10' (4 m × 3 m) office-collective design options: (A) air return located in top of drapery pocket to collect warm air for HVAC; (B) 2 × 4 air-handling fluorescent light fixture can be used for either air return or distribution; fixture will increase the ballast life. (Milton Swimmer Associates, Beverly Hills, California.)

One method of establishing light control for a typical office building (building A) is indicated in Fig. 2-7. Assume building A lighting is circuited such that rows a, b, c, d, and e each represent a separate circuit. Photocell A may have switching control over circuits d and e, while photocell B controls circuits a and b.

Observation may determine that building A segment 1 is sufficiently lit until outdoor available light falls to a given level in the afternoon, at which time photocell A turns on lighting circuits d and e. Later, at a somewhat lower outdoor light level, photocell B turns on circuit e, while, at a still lower outdoor light level, photocell C turns on circuits a and b.

The actual number and the location of photocells, as well as the rows to be controlled, are normally dependent upon actual building layout, fenestration, and orientation to the sun. A prerequisite for automatic lighting control is that lighting

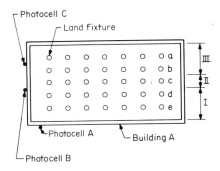

Fig. 2-7 Photocell control to maximize effectiveness and available daylighting. (Milton Meckler, "Improve Building Energy Management with Proven Electrical Conservation Techniques," *Electrical Consultant*, vol. 91, no. 11, part 1, November 1975.)

circuitry be carefully planned to allow for incremental control of portions of building lighting. Contactor control should be furnished to allow for control of several circuits by means of a single photocell. Furthermore, photocell bypass switching should be furnished to allow for maintenance of lighting circuits when required.

2-7 ASHRAE STANDARD 90-75

An important document which has proven of great help in interior and exterior design is ASHRAE Standard 90-75: "Energy Conservation in New Building Design." It was developed by the American Society of Heating, Refrigerating, and Air Conditioning Engineers and is available through the headquarters office of the *ASHRAE Journal* in New York.

ASHRAE Standard 90-75 deals with four general approaches[5] to achieving energy conservation in new building design. In secs. 4 through 99, we find both a prescriptive and a performance approach based on an element-by-element design analysis of various building materials and systems. In sec. 10, we find a systems analysis approach which outlines an alternative to secs. 4 through 9. This approach allows for compliance if the building's energy consumption is demonstrated to be equal to or lower than that obtained through application of secs. 4 through 9.

In sec. 11, we find an energy augmentation which permits the use of solar, wind, or other renewable energy source to supplement the conventional fuel sources. This augmentation is treated as a credit against the energy supplied by nonrenewable sources. Finally, sec. 12, adopted subsequently, addresses the determination of a building's energy consumption based on the energy source, as opposed to the energy supplied at the building boundary.

While we will not attempt to deal with the technical aspects of this document, Table 2-2 briefly lists some selected materials and provides data regarding their current and possible use.

In a study[9] of ASHRAE Standard 90-75, twenty prototypical building type–location projects were investigated. The Standard was found to be very effective in

TABLE 2-2 Summary of Economic Impacts due to ASHRAE 90

	Total Annual Market, millions of $	Market Affected by ASHRAE 90, millions of $ (%)	Maximum Potential Impact by ASHRAE 90, millions of $	Percent of Total Market	Percent of Affected Market
Building Materials Suppliers					
Insulation:					
Batt	1000	595 (60)	+179	+18	+30
Rigid board	470	270 (57)	+ 45	+10	+17
Loose fill	460	280 (61)	+128	+28	+46
Siding materials	70	45 (64)	+ 6	+ 9	+13
Flat glass	1000	850 (85)	+ 12	+ 1	+ 1
Windows	1247	146 (12)	+ 7	+ 1	+ 5
	903	720 (80)	− 19	− 2	− 3
Building Equipment Manufacturers					
Electric lamps	1177	176 (15)	− 16	− 1	− 9
Lighting fixtures	1450	830 (57)	−175	−12	−21
Gas and electric meters	173	159 (92)	+ 3	+ 2	+ 2
Hot water heaters	289	117 (40)	+ 4	+ 3	+ 3
HVAC Systems Manufacturers					
HVAC equipment	2308	1,720 (75)	−185	− 8	−11
HVAC controls	550	410 (74)	+ 21	+ 4	+ 5

SOURCE: "Energy Management Marketing Information," *Air Cond. Refrig. Bus.*, **22** (January 1974).

reducing annual energy consumption in all building types. It was found to reduce energy consumption in office buildings by approximately 60 percent, in retail stores by approximately 40 percent, and in school buildings by approximately 18 percent. Furthermore, the Standard proved more effective in colder climates, since space heating requirements contributed from 60 to 75 percent of the total reduction in actual energy consumption. The prototypical building models ranged in energy-utilization index (EUI) from 67,000 to 72,000 Btu/ft^2 (761,000 to 817,600 kJ/m^2) as compared with the stated GSA goal of 55,000 Btu/ft^2 (624,600 kJ/m^2). Sections 5 and 6, which deal principally with HVAC equipment systems and controls, appear to provide the principal energy reduction measures for nonresidential buildings. In terms of economics, it was found that the Standard generally increased the cost of the exterior walls, floors, roof, and domestic hot water systems. Unit costs for HVAC equipment, distribution systems, and lighting were significantly lower and tended to compensate for the cost increases noted above.

Thus, ASHRAE Standard 90-75 modified buildings costing no more to build and having significantly lower energy costs. Even where total initial costs increase, the operating savings can make up those costs in a matter of months. When average annual energy savings were compared on a straight pay-back basis on nonresidential prototypical buildings, office buildings required only 2.5 months, schools 4.6 months, and retail stores 7.6 months for the amortization of any increased initial building cost.

A number of states are now adopting their own energy codes, which may supersede ASHRAE Standard 90-75.

DESIGN CONSIDERATIONS FOR INTERIOR DESIGN

The building designer should be familiar with the concept of annual energy budgets (AEBs). Energy flow sheets should be developed for various building occupancy time periods and design days. The time effects of changes in weather and the spatial organization of the building must be considered with care. A structure's different functions and periods of use should be taken into account. Lighting levels, the color of wall surfaces, and proper fenestration are other areas which should be evaluated by the building designer with respect to energy conservation.

SUMMARY

Opinion polls tell us that economics is the basic rationale in conserving energy; energy conservation studies (ECS), if they are to be successful, must deal with the problem of reducing energy costs. Helpful in this regard are the concepts of the annual energy budget (AEB) and the energy flow sheet.

The role of the planner-designer in the initial design of a building is vital in reducing energy consumption. Also important is the role of the electrical designer, particularly with regard to lighting.

Finally, we may find ASHRAE Standard 90-75 a helpful document in planning new buildings. It contains a prescriptive approach, a performance approach, and a systems analysis approach to the problem of energy conservation in buildings. A number of states are now adopting their own energy codes, which may supersede ASHRAE Standard 90-75. Some are variations on ASHRAE Standard 90-75. Designers should verify the applicable codes in their areas.

NOTES

1. "The Effects of Price on Energy Conservation," Report No. FEA/D-75/663, Aug. 18, 1975.
2. James I. Coggins, "An Effective Energy Audit: The Flow Sheet Approach," *ASHRAE J.* (June 1976).
3. "Evaluation of Building Characteristics Relative to Energy Consumption in Office Buildings," Report No. FEA/D-76,006, Sept. 22, 1975.
4. "Technical Options for Energy Conservation in Buildings," NBS Technical Note 789, July 1973, Foreword.
5. "Energy Conservation Design Guidelines for Office Buildings," January 1974, General Services Administration/Public Building Services.
6. "Energy Conservation Program Guide for Industry and Commerce," 1974, U.S. Department of Commerce/National Bureau of Standards—Federal Energy Administration.
7. "Energy Conservation Guidelines for Existing Office Buildings," February 1975, General Services Administration/Public Building Services.
8. M. Meckler, "Improve Building Energy Management with Proven Electrical Conservation Techniques," *Electr. Consultant,* **91**(11), pt. 1 (November 1975).
9. "Energy Conservation in New Building Design, An Impact of ASHRAE Standard 90-75," Paper 43B, FEA-76/078, National Technical Information Service.

Implementing Energy Conservation Measures

An effective energy conservation program requires the cooperation of people because it is people who cause energy to be consumed through their demand for water, heat, air conditioning, light and power.

G. S. Borek and R. F. Fairbanks

Energy conservation practices have become a key ingredient in the design of all new buildings and in the retrofitting of existing buildings. Yet our expectations for conservation must be tempered by certain realities. Eighty-five percent of all commercial buildings expected to be in operation in the year 2000 exist today. Most of this building population was designed with low priorities for thermal efficiency, justified on the basis of inexpensive and readily available energy sources.

What this means is that much emphasis must be placed on the design or modification of a building's internal systems. Care must be taken to make sure that a building's heating, cooling, lighting, and electrical systems are all designed and operated in such a way as to conserve energy.

3-1 HEAT RECLAMATION STRATEGIES

Heat reclamation techniques[1] can be a major factor in accomplishing significant reductions of overall building energy requirements for both new and existing HVAC systems.

Mechanical exhaust systems expel conditioned air from the interior of a building and simultaneously introduce outside air for ventilation purposes. This outside air must then be treated to achieve standards set for temperature and humidity. The need to completely recondition "new" outside air with each exchange represents a sizable expenditure of energy in commercial structures and especially in industrial buildings, laboratories, or hospitals where numerous air changes are required each hour. Reclaiming some of the conditioning potential of the exhaust air can prove advantageous.

Through the use of heat wheels, exhaust air can be employed to condition incoming air. Two types of heat wheels are generally available: the first transfers only sensible heat, while the second, a total heat wheel, affects latent heat as illustrated in Fig. 3-1. Each wheel assembly comprises a motor-driven wheel frame packed with a heat-absorbing material—aluminum or stainless steel mesh, or corrugated asbestos. The wheels are installed in the ventilation air system, and outside and exhaust airflows are separate.

For sensible heat recovery operation, as the wheel turns, it transfers heat continuously from the warmer airflow to the cooler. Normally, the wheel is constructed so that cross-contamination is low enough for most applications. A purging section can be incorporated to reduce contamination even further. Purging is accomplished by returning a portion of the makeup air to exhaust after it has passed through the wheel.

A total heat wheel, having both temperature and humidity reduction potential, includes a filler of lithium chloride–impregnated asbestos. Lithium chloride is a desiccant which absorbs moisture as well as heat. Use of an activated heat wheel results in more energy reclaimed under summer conditions, since conditioned exhaust air can be used to cool and dehumidify makeup air.

The use of a heat wheel is limited in that inlet and exhaust ducts must be close to one another. Another approach, called a *runaround system*, overcomes this limitation. The system employs two heat exchanges connected by a loop of pipe. One exchanger unit is installed in the exhaust duct and the other in the outside-air inlet as illustrated in Fig. 3-2. A motor-driven pump continuously circulates an antifreeze solution through the heat exchangers.

Exhaust air drawn over finned coils in the first unit warms or cools the solution, while the process is reversed in the makeup-air unit. The runaround system can be as

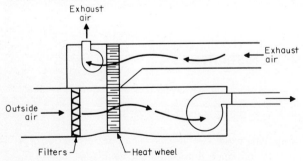

Fig. 3-1 Heat wheels. (Milton Meckler, "Heat Reclamation Strategies," *Buildings*, November 1976.)

Implementing Energy Conservation Measures 39

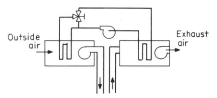

Fig. 3-2 Runaround system. (Milton Meckler, "Heat Reclamation Strategies," *Buildings*, November 1976.)

efficient in energy transfer as the heat wheel if the heat exchangers have sufficient capacity. Heat exchanger capacity can be increased by adding rows of finned tubing. However, the pressure drop through the tubing increases accordingly. The gain in efficiency, therefore, may be partly offset by the higher fan power required to move air over the coil fins.

The air-to-air heat exchanger resembles an open-ended steel box with a rectangular cross section that contains many narrow passages arranged in a cellular format as shown in Fig. 3-3. Passages carrying exhaust air alternate with those carrying makeup air. In the heating cycle, energy is transferred from the exhaust to the makeup airstreams by conduction through the passage walls.

Transfer of energy between incoming and outgoing airflows can be accomplished by banks of heat pipes running through the adjacent walls of inlet and outlet ducts. By this means, opposite ends of the heat pipes project into each airstream. The heat pipes are short lengths of copper tubing, sealed at the ends by snug-fitting cylindrical wicks and a charge of refrigerant.

A temperature difference between the ends of a pipe causes liquid in the wick to

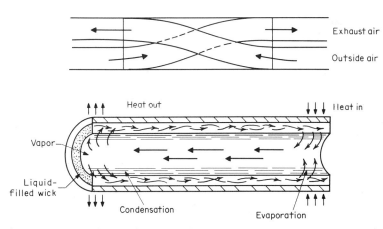

Fig. 3-3 Air-to-air heat exchanger and heat pipes. (Milton Meckler, "Heat Reclamation Strategies," *Buildings*, November 1976.)

migrate by capillary action to the warmer end where it evaporates and absorbs heat. The refrigerant vapor then returns through the hollow center of the wick to the cooler end where it gives up this heat, condenses, and repeats the cycle. Heat pipe units are often highly efficient, and, because they are sealed and have no moving parts, maintenance is minimal.

3-2 HEAT RECOVERY FROM LIGHTING SYSTEMS

There are two major methods for recovering lighting heat: plenum-return systems and water-cooled luminaires. Ducted-air systems can be designed to provide a direct means for controlling and redistributing heat dissipated by lighting fixtures. A lighting fixture provided with slots through which return air is drawn into the ceiling plenum is one approach. As heat passes over the lamps, ballasts and sheet metal of the luminaire, the air picks up as much as 80 percent of dissipated heat and carries it into the plenum.

There are many ways plenum heat can be put in use in space conditioning systems. In a double-duct or variable-volume system, the plenum can be the source of heat for air supply, with supplementary heat supplied by duct heaters or water coils piped to the space conditioning water side of a double-bundle condenser.

The use of some air-return fixtures can be beneficial when lighting heat is drawn off and discharged outdoors rather than used to heat occupied space. This approach reduces the cooling load for the building and can result in economies in ductwork distribution and refrigeration equipment. Another advantage is that, because lighting fixtures operate at lower temperatures, they produce more light for a given energy input.

3-3 PLENUM-RETURN AND HYDRONIC SYSTEMS

The two general approaches for plenum-return systems are termed the *total-return* and the *bleed-off* systems. In a total-return system of the type illustrated in Fig. 3-4, air is introduced into the room through conventional air diffusers. All air is returned through the luminaires. A fixed portion of return air is exhausted to the outside by ventilation, while the remaining air may be recycled or exhausted, depending on outdoor temperature and humidity.

Total return has the advantage of maximizing light output from fluorescent fixtures while reducing the temperature of the luminaire surfaces and minimizing bothersome radiant heating effects. This is accomplished with little change in the cooling tonnage required from that of a conventional ducted-air system.

In the bleed-off system illustrated in Fig. 3-5, most air entering a space is returned

Implementing Energy Conservation Measures 41

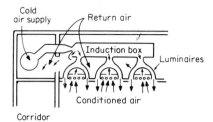

Fig. 3-4 Total return system. (Milton Meckler, "Heat Reclamation Strategies," *Buildings*, November 1976.)

to the air-handling unit directly through conventional registers. Only a portion is drawn off through the lighting fixtures and diverted directly to exhaust. Bleeding off ventilation air through the lighting fixtures generally offers the largest potential cooling capacity reduction of all air-handling methods, especially for applications that require high ventilation rates. Lighting efficiency is increased and radiant heating effects are reduced. The use of both total-return and bleed-off systems usually permits reduction in the number of air changes because of the system's direct removal of lighting heat.

Water-cooled luminaires are components of hydronic systems.[2,3] They have also proven effective for transferring heat from lighting fixtures. These units are equipped with aluminum or steel reflector housings for integral water passages as shown in Fig. 3-6. The circulating liquid absorbs heat from lamps and ballasts and carries it away. The water-jacketed fixtures transfer about 10 percent of the lighting load to the circulating liquid. In one such system, fluid is pumped through an evaporative heat exchanger to dissipate this energy and minimize internal heat gain during the cooling season. This permits the use of smaller fans, ducts, and refrigeration equipment and

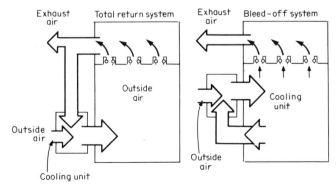

Fig. 3-5 Total return and bleed-off system. (Milton Meckler, "Heat Reclamation Strategies," *Buildings*, November 1976.)

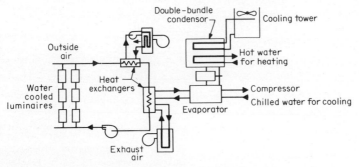

Fig. 3-6 Water-cooled luminaires. (Milton Meckler, "Heat Reclamation Strategies," *Buildings*, November 1976.)

results in lower operating costs. With this type of system, the circulating pump can be stopped during the winter to allow lighting heat to enter the space.

In more sophisticated hydronic systems circulating pumps operate year-round, and the luminaires are used in conjunction with aluminum water-filled louvers installed behind exterior glass. In the cooling season, a pump draws water from both louvers and luminaires and delivers it to a closed-circuit cooling tower. The cooling tower serves to maintain water temperature in the 75 to 85°F (23.9 to 29.4°C) range by removing excess heat and venting it to the atmosphere. The benefit of this arrangement is that it limits solar heat gain as well as gains from the lighting. During the winter, the cooling tower is isolated from the circuit and water flows from the luminaires directly to the louvers. The warm water moving through the louvers offsets heat loss through exterior glass areas and minimizes drafts.

3-4 HEAT RECOVERY FROM REFRIGERATION SYSTEMS

Three major approaches for recovering heat from refrigeration systems are conventional heat pumps, heat pump transfer systems, and the chiller double-bundle condenser.

Numerous studies[1,4] have shown that heat pumps consume less purchased energy than other, more conventional systems for heating and cooling buildings. They also incorporate into a single system, safe, year-round, pollution-free operation. Significant improvements in performance and reliability over the last several years, and the increasing cost of energy, make heat pumps viable and possibly superior to other systems, particularly when both winter heating and summer cooling are needed. It is expected that heat pumps will constitute an increasing portion of new heating and air conditioning equipment, particularly since they significantly improve load management profiles, a major concern to utility planners. Heat pumps are also especially suited to remodeling.

Further reductions in energy consumption by heat pumps are possible because of their ability to utilize low-temperature energy at the evaporator coil and convert it to higher-temperature energy suitable for heating. Suitable sources of low-temperature energy are the waste heat from power and processing plants and heat generated by equipment and lighting.

Typical examples of low-temperature heat sources include exhaust-air condenser water, well water, heat release from typewriters, computers, or other electrically operated equipment, and space lighting in the building. Proper utilization of such available heat sources can result in substantial energy savings during the heating period.[5]

Outside air has some heat content even at low temperatures. This free heat makes it possible for the heat pump to supply more energy than it consumes. Such pumps can easily attain a heating coefficient of performance (ratio of useful heat output to electric energy input) of 2 or better. The heat pump has the lowest operating cost of any electric heating-cooling equipment, provided that a heat service is available for the majority of operating hours.

Another significant source of supplemental energy source is the sun. We shall deal with this in Chap. 6.

Conventional heat pump operation may or may not involve reversal of the direction of refrigerant flow. The (reversible-cycle) heat pump is similar to a refrigeration machine and has the same basic components: compressor, condenser, and evaporator. In this type of heat pump, the direction of the refrigeration cycle is switched through operation-reversing valves as illustrated in Fig. 3-7.

An interesting alternative to the conventional heat pump is an air-source heat pump which not only transfers internal heat when it is available but can also resort to the atmosphere as a source when there is no surplus internal heat. A similar option is

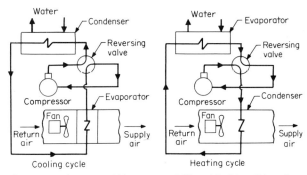

Fig. 3-7 Schematic of heat pump. (Milton Meckler, "Heat Reclamation Strategies," *Buildings*, November 1976.)

also available with water-source heat pumps. An air-to-air heat pump can be applied as a heat recovery machine in, for example, a double-duct system capable of simultaneous heating and cooling. In this type of operation, the compressor delivers hot gas to a heat exchanger (condenser) in the hot duct where the refrigerant condenses. The gas then passes through the cold duct heat exchanger (i.e., evaporator) and returns to the compressor through the suction line as shown in Fig. 3-8.

The effect of this cycle is to recover heat from the cold duct and deliver it to the hot duct. During unoccupied times when there is no surplus internal heat to be recovered, the refrigerant circuit is automatically switched from the cold duct to a roof-mounted heat exchanger (i.e., evaporator) and heat is abstracted from the atmosphere.

In cooling, the refrigerant circuit is automatically switched again, this time from the hot duct to the roof-mounted heat exchanger, which then serves as a condenser expelling heat to the atmosphere. No hydronic circuits are involved because all heat transfer is accomplished directly through the refrigerant.

Another variation, the water-to-air heat pump, can be used effectively in multizoned structures of the type illustrated in Fig. 3-9. The system is actually a hybrid configuration with some of the characteristics of a decentralized system. It also requires some remote equipment that is common to all the building zones. In this system, a heat pump unit is placed in each zone. The water-to-refrigerant heat exchangers for all these units are connected by a closed loop of circulating water. Heat is rejected into the water by heat pumps on the cooling cycle and absorbed from the water by units on the heating cycle. Also connected to the water loop at a central location are a boiler and an evaporative cooling tower, which operate as required to maintain loop water temperature between 60 and 95°F (15.5 and 35°C) year-round.

No exposure to outside air is required, so there is some flexibility in the location of such heat pumps. The system provides all the operating flexibility of three- or four-pipe systems with only two pipes. The pipes require no insulation because of the moderate water temperature in the loop. Choice of temperature in each zone is completely independent of the season or the mode of operation in other spaces.

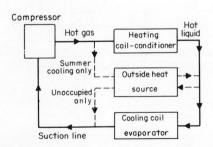

Fig. 3-8 Air-source heat pump transfer system. (Milton Meckler, "Heat Reclamation Strategies," *Buildings*, November 1976.)

Implementing Energy Conservation Measures

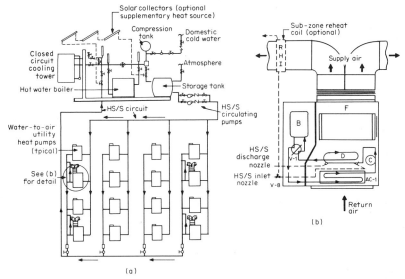

Fig. 3-9 Water-source heat pump transfer system: (a) flow diagram; (b) schematic diagram of conventional water-to-air heat pump. (Milton Meckler, "Cost Effective Solar Augmented Heat Pump/Power Building Systems," *Proceedings of the Institute of Environmental Sciences*, Los Angeles, April 7, 1977.)

Recent comparison studies indicate greater energy savings than those realized by the highly prized, variable-air-volume systems.

In Figure 3-9a, notice that the unitary water-to-air heat pumps are individually interconnected, with the heat source/sink (HS/S) closed water loop serving each of the various building temperature zones. The pumps either extract heat for space heating or discharge heat for space cooling in each such zone. Such water-to-air heat pump systems are energy-conserving only in a limited sense, since only heat from interior to exterior spaces can be transferred through the HS/S loop.

Figure 3-9b is a schematic arrangement of components within a typical commercially available water-to-air heat pump which operates as follows: With a typical zone low-voltage wall thermostat calling for heating, blower F starts and reversing valve V-1 is energized, thereby determining the refrigerant flow path. Compressor B starts and delivers hot compressed gas to the direct expansion supply air coil, and air coil AC-1 releases its heat to the recirculating air. This air, returning from the conditioned space(s), is warmed and delivered to the supply duct. Liquid refrigerant then passes through capillary restrictor tube C, causing a pressure drop, and expands (boils) into a gas within the coaxial refrigerant-to-water heat exchanger D. In this operating mode, heat is absorbed from the circulating HS/S circuit. Refrigerant gas then passes through the second path of V-1 to the suction side of compressor B, where the cycle continues. In the cooling mode, reversing valve V-1 switches the flow path between air coil AC-1, causing it to serve an evaporator and heat exchanger D, causing it to operate as a condenser, etc. Where subzone control is desired from a single unitary heat pump, heated (HS/S) water from heat exchanger D may be diverted through V-2 to air coil RH-1 as shown.

The use of solar collectors employed in conjunction with unitary water-to-air heat pumps has been explored. It is shown here as an optical heat source. Numerous parametric studies of solar-assisted unitary water-to-air heat pump systems have been undertaken to explore impacts on overall operating economies.[6,7]

Another major approach to recovering heat from refrigeration systems is the chiller double-bundle condenser, which is schematically illustrated in Fig. 3-10. If both the evaporator and condenser of a refrigeration machine are piped to appropriately located fan-coil units, the same machine can supply both cooling and heating simultaneously. This is exactly what is done in a refrigeration-type heat recovery system, although, in actual application, the refrigeration machine is equipped with a double-bundle condenser.

A double-bundle condenser is constructed with two entirely separate water circuits enclosed in the same shell. Hot refrigerant gas from the compressor is discharged into the condenser shell where its heat is absorbed by either one of the water circuits or by both simultaneously, depending on requirements of the system at a given time. One circuit is called the *building water circuit* and the other the *cooling tower circuit*. The condenser is split into two independent hydronic circuits to prevent contamination of the building water and its associated pipes, coils, pumps, and valves by cooling water which may contain dirt and corrosive chemicals.

When a double-bundle condenser is added to a standard refrigeration machine, the heat rejected by the compressor is made available to the building water circuit. Condenser water temperature, which is normally in the range of 100 to 105°F (37.8 to 40.6°C), can be boosted by adding supplementary heating, such as immersion heaters or a boiler, to the circuit. Higher water temperatures can also be obtained by

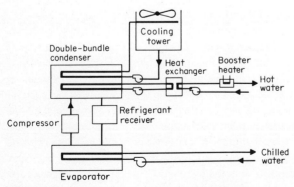

Fig. 3-10 Double-bundle condenser. (Milton Meckler, "Heat Reclamation Strategies," *Buildings*, November 1976.)

Implementing Energy Conservation Measures 47

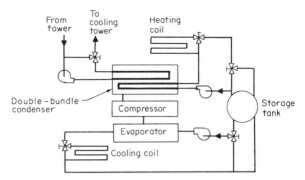

Fig. 3-11 Chillers as heat-recovery machines. (Milton Meckler, "Heat Reclamation Strategies," *Buildings*, November 1976.)

employing two liquid chillers with their hydronic circuits connected in tandem. Hot water at 150°F (65.6°C) can be obtained from the second liquid chiller, which receives 100°F (37.8°C) water from the first.

In certain heat recovery applications, the amount of heat reclaimed during occupied hours may exceed the daytime heating requirements of perimeter zones. This excess heat can be stored for release when the building is unoccupied by a system which includes a water storage tank. The temperature of the condenser water is held at about 100°F (37.8°C) by mixing storage tank water with return water as shown in Fig. 3-11. Some hot return water flows through the tank, and some is bypassed and mixed with cold water withdrawn from the tank. This continues until the tank reaches maximum design temperature, whereupon it is locked out of the circuit. If the system continues to supply an excess of reclaimed heat, condenser water is pumped to the tower for cooling.

At night, the water in the tank is pumped through the evaporator, where it serves as a "false" heating load on the refrigeration machine. When the tank temperature drops to a predetermined point, the refrigeration machine is shut down and the boiler is energized to provide hot water directly.

Another promising chiller system modification involves the operation and use of the so-called *thermocycle economizer*. Thermocycle is an energy-conserving feature that provides a means for reducing the cooling load on the central plant whenever the outdoor wet bulb temperature is 5 to 10°F (−15 to −12.22°C) lower than the chilled water design temperature. The thermocycle[7] should, under appropriate conditions of weather and building load, permit substitution of a 5-hp refrigerant pump for a 1000-hp compressor, for instance, running at light load. In most areas of the country this situation exists on many days during the year. Outdoor wet bulb temperatures are about 10 to 15°F (−12.22 to −9.44°C) lower than dry bulb temperatures.

Please examine Fig. 3-12. Assume (1) that building core areas call for cooling

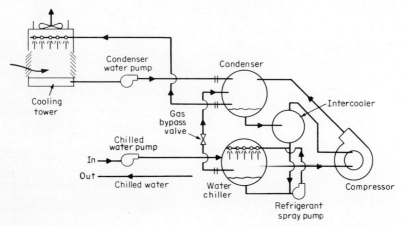

Fig. 3-12 Operation of thermocycle economizer. ("Wind-Tunnel Tests and Electric Heat Recovery System Make Glass Curtain Walls a Practical Choice for Auto Maker's New Headquarters," *ASHRAE Journal*, February 1977.)

amounting to about 25 percent of rated chiller output; (2) a chilled water temperature of 55°F (12.78°C), and an outdoor reading of 45°F (7.2°C) wet bulb temperature. Now assume that the chiller compressor is shut down while chilled water and condenser water pumps continue to run. Operating under partial load, the cooling tower can cool condenser water to within a few degrees of wet bulb temperature. The relatively warmer chilled water causes the refrigerant to evaporate, whereupon it migrates through the inoperative compressor to the lower-pressure area created by tower water in the condenser. There, the refrigerant condenses to a liquid and returns by gravity to the chiller to repeat the cycle.

In a conventional installation, this is a slow and inefficient process. Thermocycle speeds the process by adding a pipeline which allows the refrigerant gas to bypass the tight clearances of the compressor rotor. The other essentials are a refrigerant pump of 3 to 5 hp and an array of nozzles installed above the coils of the chiller. When the pump is activated, refrigerant liquid foams over the chiller coils in much the same manner as it would were the compressor in operation.

3-5 CONSERVATION OPPORTUNITIES IN LIGHTING

Significant opportunities for saving energy exist in the area of lighting. The *task-lighting* approach is recommended when it can be predetermined that a given "task" will be accomplished in a given location. An example of reduction of lighting levels in nonessential areas while providing sufficient lighting where required is illustrated in Fig. 3-13.

Implementing Energy Conservation Measures 49

The area shown consists of larger offices containing some six work stations, interspersed with aisle (circulation) spaces containing two secretarial work stations. This area may have originally been provided with a relatively high level of lighting, say, 125 foot candles (38.10 metercandles).

Assume it was found that fixtures in the large office in the *I* (interior) position could be reduced to two lamps each (from an original four-lamp configuration), while those in the *P* (perimeter) position would be left as four lamp fixtures.

The two central fixtures in the circulation space can be eliminated altogether if no work is performed in that location. Refinishing the dark walls in a light color and replacing existing dark desks can raise the coefficient of utilization (CU) by approximately 10 to 15 percent.

Retention of the four lamp fixtures near the walls is recommended to maximize wall reflectance room lighting. Resultant lighting levels can only be determined by actual footcandle measurements. It is entirely likely, however, that, for the area shown, a reduction in power usage of approximately 33 percent would lead to a lighting reduction of at best 10 to 15 percent.

While the above analysis is somewhat hypothetical, Fig. 3-14 illustrates an example of a very common wastage of energy that can easily be remedied.

Fixtures 1, 2, 3, and 4 can be fixtures containing either two, three, or four lamps, depending upon the light level originally designed for a typical office of approximately 12 by 12 ft (3.66 by 3.66 m). Little work is carried on in the vicinity of fixture 3, and none occurs near fixture 4. The room lighting, however, has originally been planned for "symmetry."

The removal of lamps from fixtures 3 and 4 leads to a direct energy savings. Fixtures 1 and 2, in the area where light is required, remain undisturbed. If fixtures in

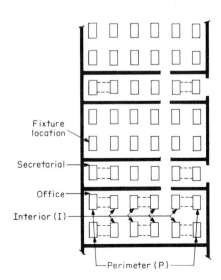

Fig. 3-13 Typical uniform general illumination lighting pattern. (Milton Meckler, "Improve Building Energy Management with Proven Electrical Conservation Techniques," *Electrical Consultant*, vol. 91, no. 11, part 1, November 1975.)

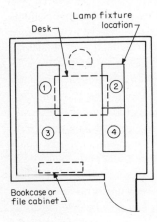

Fig. 3-14 Overlighting the small office for the sake of general uniformity. (Milton Meckler, "Improve Building Energy Management with Proven Electrical Conservation Techniques," *Electrical Consultant*, vol. 91, no. 11, part 1, November 1975.)

Fig. 3-14 are the two-lamp type, and if both lamps are removed from fixture 4, a 25 percent reduction in energy usage will result.

If, however, they are of the three-lamp type, removal of two lamps each from fixtures 3 and 4 will result in a similar 25 percent reduction. Finally, if the fixtures are the four-lamp type, the removal of two lamps each from fixtures 3 and 4 will result in a 25 percent reduction.

Redecorating the office in Fig. 3-14 in light colors will help compensate for the reduction in output for each of the conditions described above.

While it is difficult to determine an accurate overall measure of energy reduction in task-oriented lighting, a careful and judicious approach to the problem can lead to reductions of 25 percent or more in most office areas. A similar approach in laboratory areas and some industrial and storage areas will also lead to energy savings.

While the previous examples deal with modifications of existing conditions, similar reasoning can be applied to new designs. In an office area where flexibility is necessary for frequent rearrangement of furniture and office spaces, a uniform pattern of fixtures may still be installed. Lamps can be removed where they are not essential.

In a permanent installation, an initial approach of installing light fixtures only where needed is recommended. In such a case, the layout of the 12 by 12 ft (3.66 by 3.66 m) office might appear similar to Fig. 3-15.

The two-lamp layout in Fig. 3-15 can effect a 25 percent savings over an office illuminated with four lamp fixtures.

However, if the original office were illuminated with four-lamp-type fixtures, the task-lighting approach would lead to a more substantial saving, since a two-lamp fixture would likely still suffice for the incidental filing cabinet or bookcase requirements. The resulting energy savings would be approximately 35 percent.

If a greater energy saving is desired and conversion of all ceiling fixtures to the

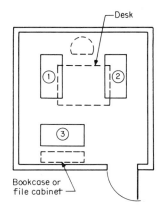

Fig. 3-15 Less lighting for needed seeing tasks. (Milton Meckler, "Improve Building Energy Management with Proven Electrical Conservation Techniques," *Electrical Consultant,* vol. 91, no. 11, part 1, November 1975.)

two-lamp variety from a four-lamp type is unfeasible because of lowered lighting levels, another alternative is available. Fixtures 1, 2, and 3 can all be two-lamp types (or four-lamp types with two lamps removed in each), while a fluorescent two-tube desk lamp furnishes supplemental desk lighting.

Many desk lamps are available with two 15-W fluorescent tubes and are generally adequate.

In this case, the energy saving over the original plan amounts to approximately 58 percent. A good light fixture maintenance program will also promote savings.

3-6 PRECAUTIONS

Care should be taken never to remove one lamp only from a two-lamp ballast circuit. Ballast failure or damage may result. The manufacturer's guarantee is also voided.

A careful switching plan for individual rooms, areas, and for the entire building is also necessary for good task lighting. An alternative approach to the removal of lamps from the light fixtures shown in Figs. 3-13 to 3-15 is to provide wall switches for turning off all unnecessary lights.

In a new facility, or in remodeling work, a careful planning of local switching can be a most significant energy conservation feature.

Providing additional switches in small offices costs little and provides a practical source for potential energy savings. Provision for additional switching and posted reminders to turn off the lights when not in use are worthwhile measures in new office planning.

Contactor controls for lighting panels can be used to shut down large blocks of lighting in occupied areas. Local switching should be utilized for controlling smaller areas. Mechanical time switches for the automatic turn off of lighting can also provide meaningful energy savings.

3-7 MORE EFFICIENT LIGHT SOURCES

The efficiency of light sources is of key importance. Incandescent filament lamps are most widely used in residential construction. They are manufactured in a variety of styles, but those generally used for residential, commercial, and some industrial applications are classified as *general-service, extended-service,* and *rough-service* types.

Table 3-1 shows some of the more common sizes and varieties of incandescent medium-base lamps available today. Types shown represent "A," "P," or "PPS" types. These are the lighting industry's designations for the most common inside-frosted bulb shapes.

Several observations can be made about Table 3-2 and its graphical representation as shown in Fig. 3-16:

1. Lamp lumens are higher for general-service lamps than for extended- or rough-service lamps in the same size range. Hence, lower-wattage, general-service lamps can replace higher-wattage lamps of the other types, with light losses kept to a minimum.

2. Light output differences range as high as 25 percent for some sizes. An average difference between the most common types and sizes used in residences (general-service and extended-service lamps in the 40- to 100-W range) is about 10 to 15 percent.

TABLE 3-1 Commonly Used Incandescent Lamps [Bulb Types "A," "P," "PS," Inside Frosted (Except as Noted), Medium Base]

	Lamp Wattage							
	25	40	60	75	100	150	200	300*
	Lumens (Lumens per Watt)							
General service†	235 (9.4)	455 (11.4)	875 (14.6)	1190 (15.9)	1750 (17.5)	2880 (19.2)	4010 (20.0)	6360 (21.2)
Extended service‡	235 (9.4)	420 (10.5)	775 (12.9)	1000 (13.3)	1430 (14.9)	2310 (15.4)	3410 (17.1)	5190 (17.3)
Rough service§	234 (9.4)		750 (10.0)	1260 (12.6)	2160 (14.4)	3400 (17.0)	5340 (17.8)	

*Clear bulb.
†Usually 750 to 1000 h, 120- or 130-V ratings.
‡Usually 2500 to 3000 h, 120- or 130-V ratings.
§Usually 1000 h, 120- or 130-V ratings.

SOURCE: M. Meckler, "Improve Building Energy Management with Proven Electrical Conservation Techniques," *Electr. Consultant,* **92**(1), pt. 2 (December 1975/January 1976).

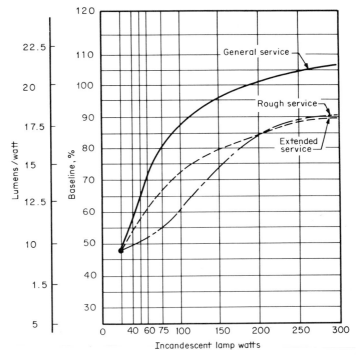

Fig. 3-16 Incandescent lamp efficiency: approximate lumens/watt versus lamp watts. (Milton Meckler, "Improve Building Energy Management with Proven Electrical Conservation Techniques," *Electrical Consultant*, vol. 92, no. 1, part 2, December 1975/January 1976.)

3. "Unit light output" or lumens per watt becomes greater with increasing lamp sizes. In other words, lamp efficiency is greater for larger sizes in nearly all cases. As a simple example, two 60-W general-service lamps provide 1750 lm, while one 100-W lamp of the same type provides the same light output.

4. General-service incandescent lamps are preferable to other types wherever they are suitable. Areas of difficult accessibility or those subject to frequent vibration or shock lend themselves to the other types.

5. A smaller quantity of higher-wattage lamps is preferable to a larger quantity of lower-wattage ones. In many cases the light output will be the same or greater, while a considerable energy saving is effected.

One common example of energy waste may be noted in restaurants, where multilamped decorative fixtures are frequently used. If a fixture of this type employs twelve 40-W lamps, for example, and these are replaced by six 60-W types, an estimated saving of approximately 25 percent should result.

TABLE 3-2 Comparison of Various Light Sources

Light Source	Nominal Lamp Watts[a]						
	40	60	75	100	110	150	175
	Initial Lumens (Lumens per Watt)						
Incandescent general service	455 (11.4)	875 (14.6)	1190 (15.9)	1750 (17.5)		2880 (19.2)	
Quartz: Recessed Single-contact PAR type						2760 (18.4)	
Fluorescent: Rapid-start	3150 (78.7)			7800[b] (78.0)	9200[b] (83.6)		
Circline	2400 (60)						
U-lamp	2900 (72.5)						
Slim-line	3000 (75)		6300				
Mercury: Self-ballasted Mogul-base				4200 (42)			8150 (46.6)
Metal halide, Mogul-base							
High-pressure sodium, mogul-base						16,000 (107)	

[a] These figures are rounded off from manufacturers' or IES handbook data.
[b] High-output lamp.
[c] Very high-output lamp.
[d] Average fluorescent lamp life generally varies over 9000–13,000 h depending on hours per start.
[e] Average life based on mean lumen rating.
[f] Rated life depends on burning position. Life may be as low as 6000 h or as high as 15,000 h.
[g] At 150 W, 12,000 h. At 400 W, 20,000 h. At 1000 W, 10,000 h.

SOURCE: M. Meckler, "Improve Building Energy Management with Proven Electrical Conservation Techniques," *Electr. Consultant,* **92**(1), pt. 2 (December 1975/January 1976).

TABLE 3-2 (Continued)

Light Source	Nominal Lamp Watts[a]							Approx. Average Lamp Life, h
	200	215	300	400	500	750	1000	
	Initial Lumens (Lumens per Watt)							
Incandescent, general service	4010 (20.0)		6360 (21.2)		10,850 (21.7)	17,040 (22.7)	23,740 (23.7)	1000
Quartz:								
Recessed	3460 (17.3)		5950 (19.8)	8250 (20.6)	10,950 (21.7)		21,400 (21.4)	2000
Single-contact PAR type					8000 (16.0)		19,400	4000
Fluorescent:								
Rapid-start		16,000[c] (74.4)						12,000[d]
Circline								
U-lamp								
Slim-line								12,000[d]
Mercury:								
Self-ballasted			7800 (26)			14,000 (18.7)		16,000
Mogul-base				22,500 (56.3)			63,000 (63)	16,000[e]
Metal halide, Mogul-base				34,000 (85)			100,000	10,000[f]
High-pressure sodium, mogul-base				50,000 (125)			130,000 (130)	[g]

If 40-W extended-service lamps were replaced by 60-W general-service types, the saving would be approximately 27 percent.

If the suggestions above were applied in incandescent lighting areas and combined with sensible zoning and switching practices, overall lighting consumption could be reduced by 15 to 25 percent.

Fluorescent lamps are far more efficient than incandescents. In those areas where fluorescents can be used, they are strongly recommended. This would apply to some residential areas, many commercial areas, and most industrial and storage areas.

56 Energy Conservation in Buildings and Industrial Plants

Certain companies have begun to manufacture a fluorescent lamp replacement for an incandescent lamp. In most cases, these consist of a ballast and Circline fluorescent lamp, with the assembly fitted with a screw base. Their use permits a significant power saving over incandescents. Cost, however, can be a prohibitive factor.

High-intensity discharge (HID) lamps include mercury, metal halide, and high-pressure sodium lamps. They are characterized by high lumen output, long life, and special starting characteristics.[8] Some of these types have been used for years in outdoor applications and in industry, particularly in high-bay-type structures. Recently a trend has developed to install them also in indoor environments. Their long life and high lumen output characteristics make them particularly attractive for large installations. This is especially true where initial installation costs and costs of relamping are high. Cold start and restart times, which run as high as 8 to 10 minutes in some cases, must be carefully considered in any installation.

Table 3-2 provides a comparison between some common light sources. Lamp lumens per watt are shown for the more commonly used light resources. This table is intended only as a general guide and not as a specific design document. Obviously, all available lamp types are not covered. Ballast or "auxiliary" losses have been neglected in discharge-type lamps for simplicity of calculation. This makes the fluorescent, mercury, metal halide, and high-pressure sodium lamps appear somewhat more efficient than they actually are.

Table 3-2 illustrates magnitude differences in terms of light output and lamp life between the major sources on the market today. More specific information is available in manufacturers' catalogs and in the latest addition of the Illuminating Engineering Society (IES) *Lighting Handbook*.

For almost all categories shown, figures for highest initial lumens and longest life are presented in Table 3-2 to establish a basis for comparison.

Figure 3-17 graphically displays the Table 3-2 data for lumens per watts versus lamp watts for the several light sources.

Finally, a word must be added about the efficiency of luminaires or complete lighting fixtures. Comparison of tables for similar types of luminaires reveals differences for CU. Just as different types of lamps should be examined for lumen output, so should luminaire photometric tables be studied.

Once a particular style of luminaire has been settled on for a given application, several manufacturers' catalogs should be checked for the highest CU (essentially representing luminaire efficiency) for that particular style.

Frequently, a change in manufacturer or a slight change in style can effect a saving of 10 to 15 percent in power requirements. For example, assume that surface-mounted industrial fluorescent light fixtures have been chosen for a typical active storage area with these dimensions and characteristics:

Room dimensions: 30 ft (9.14 m) wide × 60 ft (18.29 m) long × 10 ft 6 in (3.20 m) ceiling height.

Room cavity ratio: 3.5

Reflectance values: ceiling 80 percent; floor 30 percent; walls 50 percent
Coefficient of utilization (CU): 0.63
Footcandles required: 30

$$\text{Footcandles} = \text{fixtures} \times \frac{\text{lamps}}{\text{fixture}} \times \frac{\text{lumens}}{\text{lamp}} \times \frac{\text{CU maintenance factor (MF)}}{\text{room area}}$$

or

$$\text{Fixtures required} = \frac{\text{footcandles} \times \text{room area}}{\frac{\text{lamps}}{\text{fixture}} \times \frac{\text{lumens}}{\text{lamp}} \times \text{CU} \times \text{MF}}$$

The above formula indicates that the number of light fixtures required is in inverse proportion to both the CU and the MF. Both these quantities are variables over which the designer has some measure of control.

The other quantities are fixed for a given room and footcandle level. Assume that

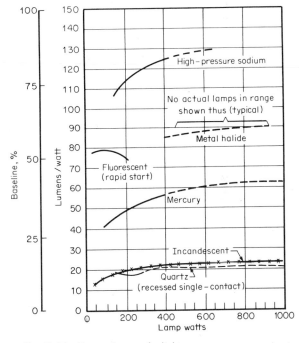

Fig. 3-17 Comparison of light sources: approximate lumens/watt versus lamp watts. (Milton Meckler, "Improve Building Energy Management with Proven Electrical Conservation Techniques," *Electrical Consultant*, vol. 92, no. 1, part 2, December 1975/January 1976.)

we are employing light fixtures with two 4-ft rapid-start cool white lamps. Under these circumstances, we would require:

$$\frac{0.63}{0.7} \times 18 = 16.2, \text{ or 16 fixtures}$$

While the number of lamps per fixture and other factors remain constant, the number of fixtures is reduced, effecting a power saving (as well as an initial cost saving) of

$$\frac{18 - 16}{18} \times 100 = 11.1 \text{ percent}$$

While the above example oversimplifies lighting design, it does point out that careful attention to luminaire efficiencies materially affects energy requirements.

3-8 LIGHTING FIXTURE MAINTENANCE

Lighting fixture maintenance by a competent, fully equipped maintenance company usually provides better results than maintenance by in-house personnel. Lens washing, fixture cleaning, and group relamping should be done on a regular basis.

In-house maintenance crews usually have difficulty accomplishing adequate maintenance because they must also handle many other tasks. Recent IES studies claim that the cleaning and relamping of light fixtures annually, as opposed to every three years, led to an 18 percent higher lighting level.

Maintenance should include regularly scheduled upkeep. If typical maintenance factors are increased by 10 percent (which appears quite feasible by regularly scheduled maintenance), then 10 percent fewer light fixtures (hence 10 percent less power) may be needed to furnish the same footcandle level.

Building maintenance practices can be wasteful and unmanageable if maintenance is carried out after normal working hours. Maintenance crews frequently turn all lights on and leave them on after leaving the premises. Maintenance during regular office hours prevents this, and permits correction of poor maintenance practices if they exist. The major advantage of having building maintenance carried out during regular working hours is that building services may be shut down promptly after the workday.

3-9 VOLTAGE-REDUCTION METHODS

Voltage reduction has been used effectively as an energy-conserving measure by some utility companies in the United States and by other countries during critical periods when fuel supplies were short.

In reducing voltage, devices are purely resistive and are directly affected by a reduction in power input, as seen from the $P = V^2/R$ relationship, where P is power supplied to the load, V is voltage at the load, and R represents load resistance.

Thus, a reduction in voltage of 10 percent causes a 19 percent reduction in power. A reduction of 5 percent would cause a 10 percent reduction to these loads. The reduction, of course, is accompanied by reduced heating and dimmer lighting for resistive heating devices and incandescent lights.

While power reduction for the type of devices mentioned is fairly easily computed, savings for induction and synchronous motors as well as discharge-type lighting and electronic equipment are somewhat harder to assess.

Motor operating characteristics will vary depending upon the percentage of voltage reduction and on the loading characteristics of the motor. In a 10 percent line voltage reduction, fully loaded motors of 25 hp or over will decrease in efficiency by about 2 percent. Meanwhile, an increase in efficiency of 1 to 2 percent will actually be realized by a motor loaded to only one-half its capacity. Though motor characteristics are affected in diverse ways by voltage reduction, power to most induction motors will not change substantially, since voltage reduction is balanced by an almost equal increase in motor full-load amperes.

The summary below indicates the approximate effect of voltage reduction on various common types of loads:

1. **Incandescent lamps:** 10 percent voltage reduction causes about 30 percent light output reduction. Lamp life increases substantially.

2. **Fluorescent lamps:** 10 percent voltage reduction results in about 10 percent loss. Lamp life is not radically changed. Lights operate satisfactorily over a fairly wide voltage range. However, if ballast minimum voltage is not maintained, the lamp goes out.

3. **Mercury lamps:** About 30 percent of light output decreases for a 10 percent line voltage decrease. Excessive low voltage results in extinguishing of arc.

4. **Electronic equipment:** Sensitive voltage increases or decreases. Much equipment is built with regulated power supplies which hold voltage stable over a fairly wide range.

5. **Capacitors:** Kilowatt output varies with the square of the impressed voltage. Power-factor correction would thus be seriously affected.

3-10 DEMAND LIMITING

The practice of limiting demand peaks by automatic or manually controlled means has been used in the past to reduce power bills. Whether loads are temporarily dropped off the line (deferred) automatically or manually, the goal is to unload the utility and the user's system at those hours of the day when demand (hence power rates) is traditionally highest. The result is a lower bill for the user and more capacity for the utility company.

Total kilowatthours are not necessarily reduced; certain loads, however, are deferred to off-peak hours when the serving utility is most easily able to cope with them. High power peaks are eliminated so the utility can serve more customers without increasing its capacity.

The manual method of demand limiting calls for the customer to furnish extensive submetering facilities so that accurate records may be made of loading and demand patterns. Alarms, set to register when demand reaches a predetermined level, may be installed on submetering equipment in a large installation. Telescopes may be set up for the viewing of meters at a great distance. Extensive submetering is, of course, a prime requisite.

Once records have been established and analyzed, the user can institute a program to temporarily disconnect certain loads when feasible, and to reschedule plant operations so that the overall load is smoothed out. In general industry, certain loads occurring during hours of peak demand might conceivably be deferrable to a second or third shift. A detailed study of operations is usually required to determine the feasibility of this approach.

For the automatic scheduling of loads, devices known as *demand-limit controllers* or *load programmers* are available. These circuits are programmed to turn deferrable loads on and off automatically in a predetermined priority sequence. Typical deferrable loads are heating loads such as electric space heating, hot water heating, and swimming pool heating.

The programming of control circuitry for loads whose prime source of power is nonelectrical allows these devices to be applied in other areas. Large boilers, for example, may be programmed on and off by control circuitry.

3-11 POWER-FACTOR CORRECTION

Power-factor correction is a highly recommended practice. The cutting of energy costs, better utilization of switchgear resulting from amperage reduction for inductive loads, voltage stabilization, the reduction of losses, and the release of wasted energy to the serving local utility are all distinct benefits of improving the system power factor.[9]

The role of power-factor correction as an ECM, however, is somewhat less important, since the reduction in actual kilowatthours of energy consumption is fairly small.

Certain power savings will be realized because of the increased efficiency of motor-driven machines. Losses in transformers and conductors will also be lower. Generally, however, power requirements for given loads will not vary substantially. The main advantage of power-factor correction is in the better utilization of switchgear and generating equipment.

Implementing Energy Conservation Measures

Data are listed below for several sizes of open-type, three-phase, 1800 r/min induction motors:

Motor Horsepower	Capacitor Rate (kvar)	Reduction in Line Current, %
200	40	6
100	20	7
50	10	7
25	5	9
10	2	13
5	2	16

The reduction line current to each motor indicated would, in effect, "unload" a motor control center (MCC) or panel having motors fed from it. When an MCC is loaded up to its full bus rating (800 A, for example), the addition of power-factor-correction capacitors can provide extra capacity by reducing the panel current draw. Frequently, when an industrial plant is running close to capacity, a comparison is made between the following alternatives:

1. Add capacity to the plant by installing new substation and distribution equipment.
2. Add capacity by correction of power factor. Often the choice of capacitors is more feasible.

At present, no energy reduction credit is given to consumers for power-factor correction of their facilities. This entire matter, however, is presently under consideration by many utility-regulating agencies, since power-factor correction, in essence, adds capacity to the serving utility as well as to the customer's facilities. The power company's ability to provide power to its customer is increased, since the required amperes per kilowatt to the load is reduced. Furthermore, the kilowatt carrying capacity of transformers is proportional to the load.

There are various methods of introducing power-factor correction to an installation. These break down into the following categories:

1. Solid connection into the system through overcurrent devices, either individually at or near each load to be corrected or at the substation, so that entire banks are permanently on the line for each substation being corrected.
2. Programmed on/off connection into the system by substation personnel.
3. Automatic connection into the system by voltage or current bias. Capacitors are automatically brought into the line as voltage drops or current increases are sensed, and dropped off with decreases in line current.

3-12 FOOTSWITCH CONTROLS

Under the general area of switching and control, we will mention here a method of energy control used with certain machines in England during World War II. There is a tendency in industry to turn on many machines and to allow them to run for long periods even though it is possible to switch them on for a particular job and off again on completion.

To eliminate this wasteful practice, foot controls were installed so that the operator had to be present to operate the machine. Though application of this measure is obviously limited to certain types of machines (lathes, drill presses, power saws, machine tools, etc.), it can lead to worthwhile savings. Power consumption savings of more than 50 percent were achieved when this measure was applied in England.

3-13 REGENERATIVE POWER TECHNIQUES

Also during World War II, American and English aircraft and automotive engine manufacturers developed a method of power regeneration which conserved fuel and energy.

Hundreds of thousands of horsepower units were being wasted in water-brake-type absorption dynamometers. Power developed in the testing of engines was being dissipated in heating water. A swinging-frame dynamometer was constructed which was electrically connected to a DC motor which was, in turn, coupled to drive a synchronous generator. Generator power then became available for powering other electrical equipment.

The explanation above is obviously a simplified one. Many mechanical and electrical problems had to be solved before the system was made practical. The system, however, was used with some success; 60 to 80 percent of the available energy was recovered in most cases.

3-14 ENERGY FLOW AND CONVERSION IN BUILDINGS

One potential savings source cited throughout the literature of energy savings is the thermostat. If the thermostat can be lowered during the winter months and raised during the summer months, significant energy savings can be realized . . . . Right? It depends. Lowering the thermostat for a few hours may actually result in no reduction in fuel consumption. Mathematical models developed for tests on residential buildings indicate that the inefficiency of HVAC equipment during reheating for short setback periods of less than eight hours will offset any consumption savings. Lowering the thermostat for long periods of time will provide significant savings regardless of the thermal efficiency of the structure or the HVAC equipment.[10]

Another paradox often encountered in energy conservation programs also relates to the various energy conversion efficiencies and their respective environmental control in buildings. Such energy conversion efficiency values are often represented[11] as the ratio of energy output available for heating or air conditioning to the energy supplied to that system. This ratio permits performance comparisons of similar building systems when the value of the energy input is known and remains constant. Generally, however, it is necessary to compare various environmental control systems employing different energy sources where conversion is exterior to the building envelope.

This problem has been addressed somewhat in sec. 12 of the ASHRAE Standard 90-75 (see Chap. 2). If we define the overall efficiency of an energy system as the output energy of its last stage in conservable form divided by the total energy input, we can isolate the various pathways by which energy sources may be supplied to buildings in terms of transfer functions or stage efficiencies, e.g., output-input, etc.

By input energy we refer to the sum of the potential energy in the incoming fuel to the first stage, and any external energy used to run the various stages.[11] Each stage of the pathways may actually have multiple inputs and outputs. However, they can be simplified somewhat as shown in Fig. 3-18 if we:

1. Combine the exploration and extraction functions.

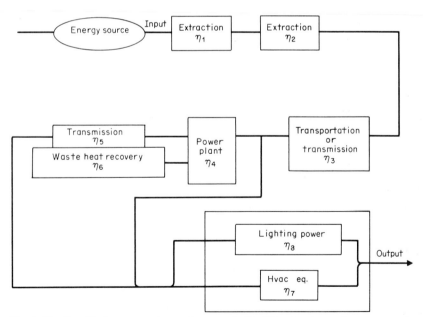

Fig. 3-18 Simplified energy pathways for control of building interior environments. (J. E. Woods and J. E. Donoso, "Energy Conversion Efficiencies for Thermal Control in Buildings," *ASHRAE Journal,* January 1977. Reprinted by permission of the American Society of Heating, Refrigerating, and Air Conditioning Engineers, Inc.)

64 Energy Conservation in Buildings and Industrial Plants

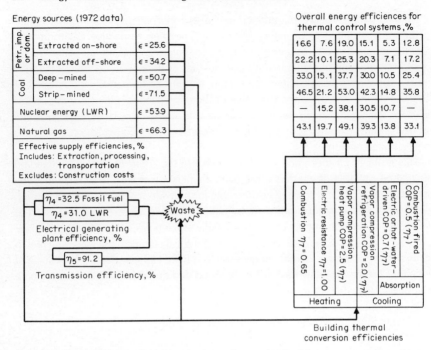

Fig. 3-19 Overall energy efficiencies for thermal control in buildings. (J. E. Woods and J. E. Donoso, "Energy Conversion Efficiencies for Thermal Control in Buildings," *ASHRAE Journal*, January 1977. Reprinted by permission of the American Society of Heating, Refrigerating, and Air Conditioning Engineers, Inc.)

2. Incorporate preprocessing transportation in the processing transfer function.
3. Combine storage with the transportation or transmission stage.
4. Neglect the energy required for construction (see Chap. 1).

We can further simplify the problem by expressing the stage efficiencies for extraction, processing, and transportation as effective energy supply efficiencies similar in format to Fig. 3-19.

3-15 CONSERVATION POTENTIALS OF AUTOMATED ENERGY MANAGEMENT SYSTEMS

Energy and cost savings from automated energy management systems fall into three main categories:

Implementing Energy Conservation Measures 65

1. More efficient operation of heating and air conditioning equipment.
2. Programmed operation of lighting to run lamps on and off when needed.
3. Selective shedding of the electrical load to avoid demand peaks.

To provide more efficient heating and air conditioning, equipment must be operated much closer to optimum than is possible with manual or local control. HVAC systems can be programmed where appropriate: Use outdoor air for cooling when the weather conditions and the time are right. Cycling boilers and refrigeration chillers in a pattern responsive to time of day and prevailing weather prevents the waste of standby losses on an excess of heating and cooling capacity during operating hours. Loading chillers to permit maximum efficiency takes advantage of the thermal lag in the cooling system and building structure.

When one or more buildings are of such size that they warrant a computer in the building automation system, software programs can be provided that integrate weather conditions, building and system characteristics, and HVAC operating conditions. Software can record the operating experience of a building, along with weather conditions, so that the HVAC system operation can be reprogrammed for greater efficiency.

Changeover from heating to cooling, and vice versa, in intermediate seasons can also be accomplished optimally by automatic monitoring of relevant conditions rather than by manual activation.

When lighting is activated by an automated energy management system, lighting systems can be turned off or on according to a predetermined schedule. Fresh-air quantities can often be reduced when the amount normally provided exceeds code requirements. Fans can be shut down sequentially for short periods.

If perimeter lighting is separately circuited, perimeter systems can be turned off automatically when daylight levels become sufficient. They can be turned off during lunch hour periods in office buildings or recesses in schools. The circuiting can also anticipate cleaning personnel needs.

Elevators can also be programmed for selective reduction, though normal elevator programming in larger buildings takes into account changing usage during the day. Sometimes programming is even based upon the traffic rate.

As we discussed earlier, building electrical systems also can be programmed on a priority basis to shut down to avoid demand peaks.

By one or more of these means, operating procedures and associated system programming can be optimized to achieve remarkable savings in energy and operating costs. Refer to Table 3-3, which indicates the savings possible for a representative 1-million-ft^2 office complex.[12] Energy cost was assumed to be 1 cent/kwh.

Building automation systems, when properly planned, can help ensure the success of the implemented energy conservation measures (ECMs). They can provide valuable operator training and can help improve maintenance practices[13] and failure detection mechanisms.

TABLE 3-3 Possible Savings for a Representative 1-million-ft² Office Complex

Operating Technique	Nature of Savings	Annual Savings, $					
		Heating Energy		Cooling Energy		Other Energy	Non-Energy
		Steam	Electric	Steam	Electric		
1. Energy conservation based on outdoor air	Normal occupancy Off-hour	12,000 5,000	35,000 15,000	7000 3000	6000 3000	2000 1000	
2. Start/stop schedule modification	Energy for motors Heating, cooling energy Equipment life	8,500	25,000	1500	1500	3000 500	1000
3. Refrigeration plant efficiency	Improved chiller efficiency Operating-time reductions			3500 2000	3500 2000	500	
4. Control of chilled water distribution	Electricity					500	
5. Energy conservation based on chilled water temperature	Cooling energy Pumping energy			1000	1000	(700)	
6. Electrical demand forecasting; limiting	Electrical demand charges						500

SOURCE: "Basics of Building Automation," NECA Electrical Design Guidelines, National Electrical Contractors Association, 1973.

Implementing Energy Conservation Measures

Once the ECMs are identified, failure to recognize the special needs of operator personnel can be crucial. These employees must successfully operate, maintain, and service the equipment. They must reduce energy consumption and be sensitive to new or overlooked ECM opportunities. The following steps have proved helpful in improving operator sensitivity and cooperation:

1. Arrange for prompt posting of charts or bar graphs that indicate monthly energy consumption and demand levels. Care should be taken to reflect operator corrective actions directly. This stimulates personal interest in further participation.

2. Arrange for water treatment contractors who furnish chemicals for chilled, hot, or condenser water systems to periodically monitor performance and advise operators on the operation of these critical systems. (Slight increases in the fouling of equipment heat transfer surfaces can result in significant increases in related system energy consumption.)

3. Independently verify that established maintenance schedules are maintained by operators. Inspect critical components, heat transfer surfaces, etc. The monitoring of head pressure at frequent intervals will point up possible problems and permit a prompt cleaning operation or elimination of noncondensables.

4. Arrange for operators to attend periodic training courses on topics such as heat transfer and energy management fundamentals.

5. Remind operators to be wary of leaks in *all* steam lines, valves, or filters and pneumatic control piping. Where possible increase pneumatic tank pressure range to eliminate excessive short cycling.

6. Actively solicit operator comments and ideas for specific ways of reducing energy consumption. Reward workers personally and publicly when implementation improves performance.

7. Have operators verify that the power factor (particularly where there is a high percentage of induction-type motors) is within reasonable limits. Correct low values to minimize operational deficiencies.

SUMMARY

Major conservation opportunities are present in the design and operation of a building's heating, cooling, lighting, and electrical systems.

Heat reclamation strategies involving heat wheels and heat exchangers can accomplish significant reductions in overall building energy requirements. Heat can also be recovered from lighting systems, through the use of plenum-return systems and water-cooled luminaires. Heat may be recovered from refrigeration systems through the use of conventional heat pumps, heat pump transfer systems, or the chiller double-bundle condenser.

Approaches such as task lighting and lamp removal can produce significant savings in the area of lighting. Careful placement of light switches, the judicious choice of lamp types, and proper maintenance can also reduce consumption levels.

Energy may be saved in electrical systems through voltage-reduction methods, the practice of demand limiting, and power-factor correction. Other options include the installation of foot-switch controls on machinery and regenerative power techniques.

A proper understanding of the role of the thermostat and a knowledge of energy conversion efficiencies will aid in avoiding some common misconceptions which tend to reduce savings.

Finally, automated energy management systems can help produce energy and cost savings in heating, air conditioning, lighting, and electrical systems.

NOTES

1. M. Meckler, "Heat Reclamation Strategies," *Buildings* (November 1976).
2. M. Meckler, G. Meckler, "Design and Economic Evaluation of Dynamically Integrated Lighting Systems," *Illum. Eng.* (February 1963).
3. M. Meckler, G. Meckler, A. Hoertz, "Dynamically Integrated Lighting and Air Conditioning," *Heat./Piping/Air Cond.* (April 1963).
4. M. Meckler, "Cost Effective Solar Augmented Heat Pump/Power Building Systems," *Proc. Ins. Environ. Sci.* (April 1977).
5. M. Meckler, G. Meckler, "Nomograph Computer Methods—Optimization of Building Systems of More Than Three Variables, Methods of Building Cost Analysis," Building Research Institute Publication 1002, 1962.
6. M. Ucar, E. E. Druckar, J. E. LaCraff, W. H. Card, "Thermal Simulation of a Building With Solar Assisted Closed Liquid Loop Unitary Heat Pumps," delivered at the winter ASME annual meeting, New York, December 1976.
7. "Solar Energy Heat Pump Systems for Heating And Cooling Buildings," ERDA Doc. Coo-2560-1, Con. 7506130 Pennsylvania State University.
8. M. Meckler, "Improve Building Energy Management with Proven Electrical Conservation Techniques," *Electr. Consultant*, **92**(1), pt. 2 (December 1975/January 1976).
9. M. Meckler, "How to Improve Building Energy Management," *Electr. Consultant*, **92**(2) (February/March 1976).
10. M. P. Zabinski, J. Y. Parlange, "Thermostat Down Fuel Consumption Up, A Paradox Explained," *ASHRAE J.* (January 1977).
11. J. E. Woods, J. E. Donoso, "Energy Conversion Efficiencies for Thermal Control in Buildings," *ASHRAE J.* (January 1977).
12. "Basics of Building Automation," NECA Electrical Design Guidelines, National Electrical Contractors Association, 1973.
13. M. Meckler, "PM Checklist," *Heat./Piping/Air Cond.* (April 1972).

Role of the Computer and Microprocessor in Energy Management

The current energy situation is having an ever-worsening financial impact on building operational budgets. More and more owners/operators are turning to their consultants with a plea "to do something . . . and do it quickly" to establish an effective energy management tool.

Gregory P. LaRocca

The growing complexity of mechanical and electrical building systems, combined with an increased vulnerability to fire and security breaches, has made computer, microprocessor, and automated energy management systems essential in more and more buildings. Buildings have grown larger and more complex. As a result, the need for automatic centralized building control and optimized energy management has grown as well.

The notion of controlling building systems from a central location emerged in the fifties. In its early days, centralized control was restricted mainly to the monitoring and operation of heating and air conditioning systems. Now, in the more elaborate systems, the entire building has a single nerve center; fire alarm, sprinkler, supervisory, and security systems are integrated; equipment monitoring, audio communications, and command systems are basic elements, as in HVAC control. Closed-circuit TV is an ancillary function.

4-1 COMPUTERIZED POWER MANAGEMENT SYSTEMS

In response to the high electrical power demand charges, several major computer hardware manufacturers have made available sophisticated microprocessors and ready-to-use software packages which transform their real-time control computers into rather elaborate load-shedding and energy optimization centers.

4-2 ENERGY MANAGEMENT SYSTEMS

Because energy management systems require a mix of specialized hardware and software, early planning is recommended. Control-monitoring flexibility and expanding capacity capability should be considered at the onset. The decision of whether to build upon hardware or software logic should reflect the designer's judgment of the need for future adaption or modification. Consider the following factors for retrofitted HVAC systems:

1. Access values of various degrees of report generation or operation intervention. Are they essential? What management load is imposed?
2. Is it desirable to check out the programmed control philosophy under simulated (versus actual) operating conditions?
3. Where automatic start/stop of HVAC equipment is scheduled with time-of-day criteria, what should the minimum cycle times be?
4. Is it desirable to modify control philosophy on the on-line mode?
5. How useful are the vendors' "ready-to-use" software packages?
6. Is it appropriate to have time-variable-demand target values–schedules?

Once these decisions are made, different vendor systems become easier to compare.

A new generation of computer-based data acquisition and control systems designed for applications beyond the scope of other building automatic systems are now becoming available. In addition to the usual functions of scanning, alarm reporting, set-point control, etc., they greatly enhance the usefulness of the existing system by allowing the user to:

1. Monitor and measure the distribution of all energy types.
2. Identify the source of inefficient energy consumption.
3. Develop algorithms for controlling and optimizing this consumption.
4. Modify these algorithms when necessary.

The ability to perform these functions requires a high-level language compiler with an easily programmable format, i.e., BASIC. In some cases, these systems can also be used for developing related and nonrelated programs such as inventory control, labor distribution, and payroll.

With the incorporation of a computer, a building automation system can determine boiler and chiller loads and running times, then calculate energy consumed, daily cost, and efficiency. And, remarkably, computer programs called *optimizing packages* are available that actually improve building operation through "learning experience." Least-cost equations based upon building history, modified by the monitoring of equipment, automatically adjust equipment start/stop schedules or system outputs for minimum cost operation. One type of computer program integrates past and predicted electrical demand. As predicted power peaks approach, the computer reschedules equipment utilization, shutting down low-priority systems until the potential overload has passed. Other programs integrate outdoor and indoor temperatures, solar load, and indoor and outdoor humidities for controlling coil temperatures, dampers, and fans in air-handling units. These programs can adjust the start-up of heating or cooling plants automatically to respond to weather conditions. They will also vary operation for occupied and unoccupied conditions.

Manufacturers of building automation systems generally agree that the following four levels of increasing sophistication can be used to discriminate between system capabilities:

1. Central monitoring and control of on/off equipment, permitting an operator to read or change the state of any point and to receive an alarm when equipment starts up or shuts off uncommanded.
2. Provision for the control system to read analog values such as temperature and pressure, adjust analog settings and damper positions, and permit analog alarming and clock programming.
3. Ease of expansion of capacity.
4. Provision for a computer. Some control systems permit this feature as an add-on, while other more elaborate control systems are computer-based initially.

In its greatest stage of sophistication, a computerized control center can perform a large number of tasks, among which are:

1. **Chillers:** Start/stop; monitor characteristics.
2. **Pumps:** Start/stop; supervise flow and motor operation.
3. **Electrical distribution:** Monitor characteristic.
4. **Cooling towers:** Control and monitor temperature and water flow.
5. **Air-handling units:** Start/stop; modulate dampers; check filter conditions; modulate valves.
6. **Lighting:** Turn on or off according to a preprogrammed schedule.
7. **Security:** Scan and interrogate protection network for alarm and status information; lock or unlock gates; provide patrol tour monitoring; read personnel identification cards.

These highly sophisticated building automation systems generally include several basic components: an operator's console with a keyboard for system access and visual display; automatic printers that provide alarm summaries, multiple-point trends, and point logs; and central processing units (CPUs) and/or minicomputers that allow for psychometric calculations, systems optimization, equipment efficiency calculations, electric utility profiles, maintenance scheduling, etc. Such a system is

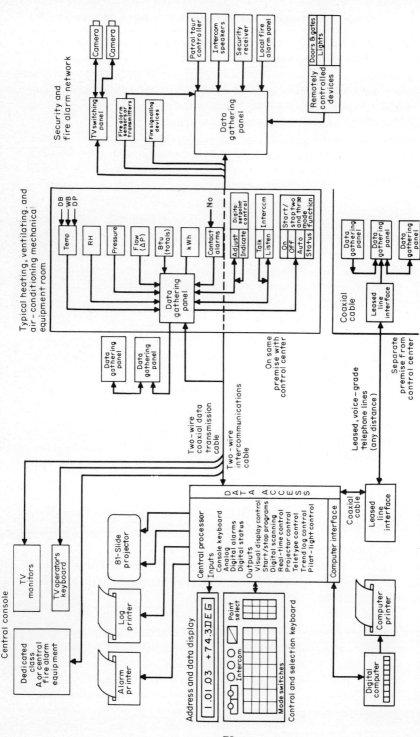

Fig. 4-1 Interrelationships of the multiplicity of building automation system devices. ("Basics of Building Automation: Design Guidelines," National Electrical Contractors Association, December 1973, p. 74.)

illustrated in Fig. 4-1. Cathode-ray tubes (CRTs) are available for alphanumeric display to indicate the results of system interrogation.

4-3 ESTABLISHING LEVELS OF CONTROL FOR ENERGY SAVINGS

There are three levels of control in any type of building automation control system: local control, central manual control, and central automatic control.

Local control systems provide independent, relatively low-cost control for specified systems and equipment. Each local controller operates independently and does not act in conjunction with any other controlling device. Wiring and communications costs are minimized, since local controllers don't require trunk line wiring as do central control systems. A local control system will entail a higher control system maintenance cost than a central control system. Refer to Fig. 4-2 for a schematic diagram illustrating local control adapted to a single-zone chilled water control system.

A central manual control system is a centralized system capable of being operated from a single command console. A trained operator is responsible for the efficient operation of the total system. This type of system is totally dependent upon the operator's decision-making capability. The effectiveness of this system also depends on the size and sophistication of the system being controlled.

The basic central manual controller is capable of status, analog, and control-point alarm indication, remote analog set-point adjustment, and remote start/stop. Status and analog indication at the central console provide information on building condi-

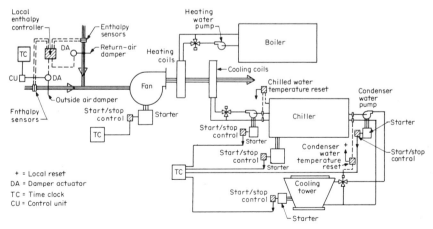

Fig. 4-2 Single-zone chilled water system with local control. (Edward R. Berger and Edwin F. Coxe, *Automation and Centralization of Facilities Monitoring and Control Systems*, ED 7601, U.S. Army Facilities Engineering Support Agency, Fort Belvoir, Virginia, May 1976.)

tions and notify the operator of abnormal conditions. The remote start/stop function allows the operator to control distant equipment from the central console. It thereby provides a prime savings by reducing work force requirements. For small, less sophisticated systems, the operator can perform optimization functions by manual inspection and control of specific remote control points. Refer to Fig. 4-3 for a schematic diagram illustrating central manual control of the same single-zone chilled water system illustrated in Fig. 4-2.

A central automatic control system (ACS) is a centraltzed system where control is accomplished by software logic or hard-wired logic systems. Software logic systems, or minicomputer-based systems, provide versatility in programming, since reprogramming can be accomplished without physical hardware changes. Hard-wired logic systems, or microcomputer-based systems, are physically wired to perform functions and cannot be changed without rewiring.

Three basic central automatic control functions covered in this study are remote start/stop, optimization, and maintenance. Remote start/stop is useful for energy conservation and labor reduction. Start/stop operations can be initiated by the operator at the central console or by programming devices such as a minicomputer or microcomputer. System optimization increases the system or apparatus efficiencies to eliminate energy losses and decrease the cost of system operation. Optimization is achieved automatically with the use of programmed devices; it requires no operator intervention. Examples of optimization programs are:

1. Start/stop optimization
2. Enthalpy optimization
3. Chiller optimization

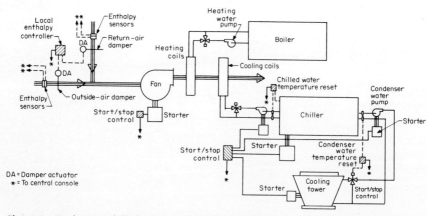

Fig. 4-3 Single-zone chilled water system with central manual control. (Edward R. Berger and Edwin F. Coxe, *Automation and Centralization of Facilities Monitoring and Control Systems*, ED 7601, U.S. Army Facilities Engineering Support Agency, Fort Belvoir, Virginia, May 1976.)

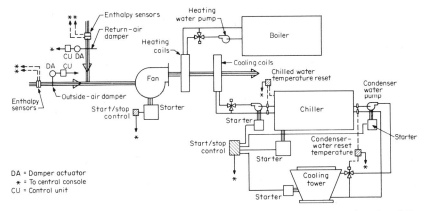

Fig. 4-4 Single-zone chilled water system with central automatic control. (Edward R. Berger and Edwin F. Coxe, *Automation and Centralization of Facilities Monitoring and Control Systems,* ED 7601, U.S. Army Facilities Engineering Support Agency, Fort Belvoir, Virginia, May 1976.)

4. Solar heat optimization
5. Fan system optimization
6. Heating-cooling optimization
7. Lighting optimization
8. Demand optimization

The system maintenance functions increase the life of equipment, thereby reducing equipment and labor costs. Although an ACS cannot actually perform maintenance, it can be programmed to provide information for maintaining the equipment more effectively. See Fig. 4-4 for a schematic diagram of the single-zone chilled water system modified to include central automatic control.

4-4 STANDARD COMPUTER OPTIMIZATION PROGRAMS

A variety of standard computer optimization programs are available for the control of building systems. Figures 4-5 through 4-12 schematically illustrate eight of the most widely used programs commercially available as programmable software or hard-wired from major control manufacturers.

The enthalpy optimization program in Fig. 4-5 minimizes the expenditure of cooling or heating energy. It compares the heat content of outdoor and/or return air and selects amounts of each to obtain the warmest air for heating and the coolest air for cooling. Each air-handling unit requires the following inputs: outdoor-air dry bulb

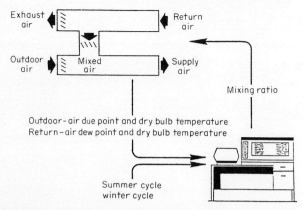

Fig. 4-5 Enthalpy optimization. (Welton Becket and Associates, Architects and Engineers, Los Angeles, California.)

temperature, outdoor-air dewpoint temperature, return-air dry bulb temperature, return-air dewpoint temperature, and the desired leaving-air temperature conditions for the cooling coil.

This program operates for either summer or winter cycles. The object is to take advantage of the heat content of the outside air. The system calculates the optimum mixing ratio based on the enthalpy of the outside, return air, and the mode of operation. Consideration is given to the costs of sensible heat, latent heat, and dehumidification.

Where there is less total heat content in the outside air in spite of the dry bulb temperature being higher than the return-air temperature, outside air is used directly to reduce the cooling load. For facilities that have high, year-round internal heat gains requiring constant cooling, the so-called "free" cooling cycle can be used to satisfy the demand, while less energy is consumed by the mechanical refrigeration equipment. Where novel situations arise, procedures[1] are available for simulating the performance of components and systems. Savings vary from 2 to 12 percent depending upon climatic conditions, hours of operation, and the conventional economizer switchover temperature normally used. This switchover is controlled only by dry bulb outside temperatures.

The variable start-time selection program illustrated in Fig. 4-6 delays the start of primary equipment, such as chillers, boilers and air-handling units, until the last possible moment. Start times are based on the space loads and external weather conditions.

Weather data, interior conditions, occupancy times, and equipment capabilities are fed into the computer. The computer determines optimum equipment start times to bring environmental areas to correct comfort conditions prior to occupancy time, with minimum machine running time and lowest possible energy costs.

Required inputs include inside temperature of the representative space for each of the air-handling systems, outside dry bulb temperature, supply and return chilled water temperature, supply and return hot water temperature, status of the nighttime fan, and the wall temperature (to account for the "cold-wall" effect).

Energy conservation is accomplished by not operating the equipment for any more time than necessary. Energy savings due to the reduction of heat loss (or gain) from the building are also obtained, since the temperature difference ΔT between the building and outdoors is maintained for a shorter duration.

A separate start of each air-handling unit, according to the need of the zones served by it, makes further savings possible. Furthermore, the hot water circuit should be maintained at the highest allowable temperature to minimize the start-up time; the temperature reset of the hot water, based on the outside air, is inhibited during the start-up time until the spaces come into condition. This action helps to reduce the time required to bring the building to comfort conditions.

The start-time optimization is accomplished with the minimum amount of cumbersome instrumentation if the calculating capability of the computer is used extensively. An updating of the algorithm used in the two subroutines (identification and minimum time) can be easily accommodated after the system has been in operation for some time and adequate experience is gained.

Where local codes permit, equipment should be started automatically. Otherwise, instructions should be printed out for the operator to start the equipment according to the optimal times calculated by the computer.

The chilled water optimization program illustrated in Fig. 4-7 applies to systems that utilize parallel-piped, single-speed centrifugal compressors. The program assumes that each compressor has its own chilled water pump and condenser pump, and that these pumps are started when the compressor starts and are shut down when

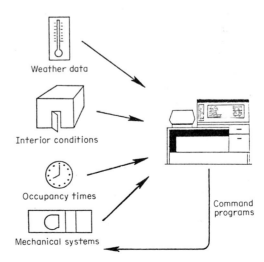

Fig. 4-6 Variable start-time selection. (Welton Becket and Associates, Architects and Engineers, Los Angeles, California.)

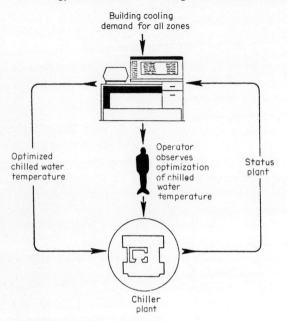

Fig. 4-7 Chilled-water optimization. (Welton Becket and Associates, Architects and Engineers, Los Angeles, California.)

the compressor is not running. Multiple-cell cooling towers, with fans that can be started and stopped, are also assumed.

Among the number of operating alternatives are varying condenser water temperatures to reduce power consumption on chillers, sequencing chillers within their maximum efficiency ranges, arranging chillers for either parallel or series flow, utilizing elevated chilled water temperatures during times when humidity conditions permit, and using heat exchangers to recover heat from the condenser water.

The program measures and predicts optimum chilled water temperature for chilled water systems. The system first examines zone cooling loads and determines the optimized chilled water temperature controller setting.

Savings calculations are part of the basic program. The calculations consider the operation of multiple compressors at maximum efficiency and the added efficiency of decreasing condenser water temperatures. Additional savings can be obtained by increasing chiller discharge water temperature.

The electrical demand load-leveling program illustrated in Fig. 4-8 assures that electrical energy usage is monitored by the power company. Monthly charges are based on total energy usage for the month plus a penalty charge for the highest kilowatt usage during any single time interval. The interval may be the last month, quarter, or year, depending on the local power company. The electrical demand load-leveling program prevents or at least lessens maximum peak energy usage.

Demand load limiting provides continuous monitoring of electrical usages among the building's systems. Interruptible loads are sequenced off for periods of time to reduce the maximum demand on the facility. If the projected energy demand indicates a total usage in excess of a preestablished demand limit, one or more of several actions can be taken. The system may print out the demand for each period, print out a demand alarm, print out interruptible loads, or initiate the automatic load.

At the end of the time interval, the power company measurement device will produce a signal to reset itself. This same signal is used to reset the program so that it can begin totalizing and estimating for the next interval. The program has an internal self-timer which gives an alarm signal or printout if this reset signal does not come within the scheduled time. This prevents the energy totalization from continuing unnoticed into an interval in which it was not used.

Electrical demand load-leveling programs normally produce savings through lower electric bills. Additional savings can be realized at the time of initial installation through appropriately sized electrical supply facilities. At present, these facilities must be sized for a specific power draw level. With the electrical demand load-leveling program, this power draw level can be lowered to some other value. The entire power distribution system can thereby be reduced in power-handling capacity.

The electrical demand load-leveling program illustrated in Fig. 4-8 assimilates weather conditions. The status of all building mechanical systems is fed into the computer. The system then responds with a demand profile enabling the operator to predict demand peaks and conduct load shedding to minimize energy costs.

The electrical demand load-shedding program illustrated in Fig. 4-9 continuously monitors demand and conducts load shedding *automatically.*

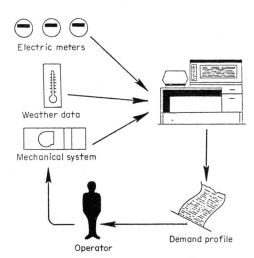

Fig. 4-8 Electric-demand load leveling. (Welton Becket and Associates, Architects and Engineers, Los Angeles, California.)

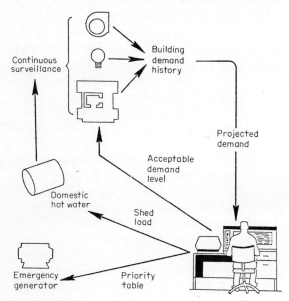

Fig. 4-9 Electric-demand load shedding. (Welton Becket and Associates, Architects and Engineers, Los Angeles, California.)

The load profiles program illustrated in Fig. 4-10 computes and integrates all necessary building information for cost reduction and long-range data analysis. The most common variables are daily and monthly cooling and heating loads, daily and monthly kilowatthours, kilowatt demand and total electricity, gas, steam or other prime energy costs.

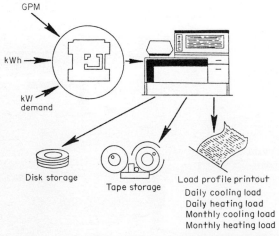

Fig. 4-10 Load profiles. (Welton Becket and Associates, Architects and Engineers, Los Angeles, California.)

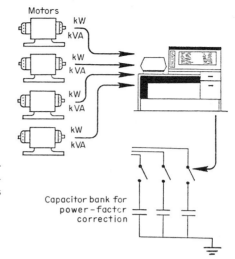

Fig. 4-11 Power-factor correction. (Welton Becket and Associates, Architects and Engineers, Los Angeles, California.)

The power-factor correction program illustrated in Fig. 4-11 continuously monitors the power factor for individual buildings or an entire complex. When the power factor drops below a present limit, the computer automatically switches in a bank of capacitors to provide power-factor correction. If the power factor is still below its desired value, the computer will automatically switch in a second bank, and so on.

The chiller-loading optimization program illustrated in Fig. 4-12 measures the

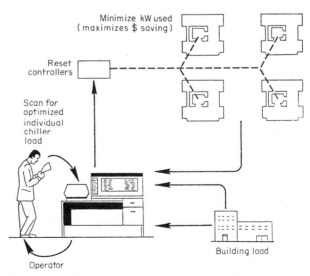

Fig. 4-12 Chiller loading optimization. (Welton Becket and Associates, Architects and Engineers, Los Angeles, California.)

total cooling requirements of a building and computes the optimized individual chiller loads so that the total energy consumption of the chillers is minimized.

4-5 CANDIDATE BUILDING SELECTION FACTORS

4-5-1 Selection Factors

Four major factors may be identified which affect the energy savings and energy control system cost potential of most buildings. These factors should be examined before deciding which buildings are suitable for control purposes. They are:

1. Building size
2. Usage patterns
3. Heating and cooling systems
4. Relative building location

4-5-2 Building Size

Building size is usually the most significant factor in the selection process. The greater the total floor area, the greater the energy consumption, except in industrial warehouses and shop buildings not centrally heated or cooled. In such structures, supervisor offices or small percentages of the floor area may be cooled by window units or small, packaged direct-expansion systems. Work areas may be ventilated during summer and heated in winter by unit heaters.

Air temperatures are not as closely maintained as in an office complex. Therefore, energy consumption by the heating and cooling system will be relatively small compared to the total floor area. Heavy energy consumption in a machine shop, for example, will probably be due to process equipment rather than the heating and cooling system. In general, the larger buildings with large HVAC systems are the most attractive for connection to centralized control systems. In a study of Fort Belvoir, Virginia,[2] buildings with more than 20,000 ft^2 proved the most attractive for connection to a central system. Recent studies indicate that buildings in excess of 20,000 ft^2 are good candidates for interconnection with energy management systems. Smaller buildings having a variety of HVAC systems throughout the structure may be more suited to local control.

4-5-3 Usage Pattern

Patterns of building usage affect energy consumption to a large extent. Recreational centers with heavy occupancy tend to be the greatest consumers of energy. They are followed by office and residential buildings. Most buildings can be loosely categorized into one of three occupancy types:

1. Evening occupancy generally involves recreational buildings such as bowling alleys and theaters. Time range is approximately 5 p.m. to 12 midnight.
2. Office occupancy is confined to buildings used during the typical workday. Time range is approximately 7:30 a.m. to 4:30 p.m.
3. Residential occupancy is essentially late evening and early morning. Time range is approximately 6 p.m. to 6 a.m.

4-5-4 Heating and Cooling System

The type of heating and cooling system can influence the energy savings potential and the expense of conversion to a central control system. Generally, central heating and cooling systems are found in larger, more complex buildings. Such structures usually produce the greatest savings potential. System components such as air-handling units, chillers, and boilers are usually located in the same area of the building, thus simplifying control system modifications.

Smaller buildings tend to have widely distributed window units and unit heaters. These are difficult and expensive to adapt to a central energy control system. Since smaller buildings tend to have small savings potential, the expense may not be justified. In such cases, a system of local control may be cost-effective.

4-5-5 Relative Building Locations

A major factor in central control system costs is wiring and communication expense. This interconnection cost is a function of building location relative to its remote panel and to the central console. Telephone lines can be used between the remote panel and the central console where suitable voice-quality lines exist. Where spare lines do not exist, installing additional telephone link capacity must be included in the system cost estimate.

The connection between the remote points and the remote panel does represent an expense. Generally, the larger energy-using facilities require many points at considerable distance from other facilities. For these reasons, many office and recreational buildings have individual remote panels.

Closely grouped buildings offer the best possibilities for controlling several facilities through one remote panel, as illustrated in Fig. 4-13. Multiple building connection to one remote panel is usually less expensive than installing a remote panel in each structure, despite increased wiring costs.

A number of monitoring points can provide the dispatcher of an energy management control system (EMCS) located at a console (see Fig. 4-14) with readily available data on the major operating parameters of each system controlled. Data will

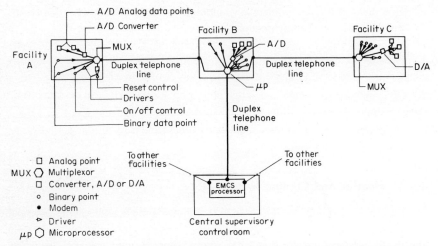

Fig. 4-13 Schematic of representative multiplexed control system. For clarity's sake intercom lines are not shown. (The Energy Group, Consultants, Los Angeles, California.)

also be provided on out-of-limit points which will generate an alarm when values outside the permissible range are reached.

Monitoring will also improve the scheduling of maintenance by providing an overall picture of maintenance required for one or more buildings.

The EMCS communicates with the equipment being monitored, receiving information as to status or analog values, and issuing analog or binary commands. Each of the data-gathering or command functions is referred to as a *point*. The points are those commonly monitored and controlled in building systems. They are listed in Table 4-1. The points are communicated as follows:

Data from Remote Equipment to Central Console

Binary   If the information can be conveyed as two states, such as an open or a closed switch contact, or the presence or absence of a voltage or a pressure

Analog   If the quantity is continuously variable, such as a voltage, a pressure, or a temperature

Command from Central Console to Remote Equipment

Start/Stop   If the command has only two possible states such as open/close, run/stop, or on/off

Reset   If the execution of the command results in a change in a continuous variable such as a voltage, a current, or a temperature setting

In local control systems, where control-point–to–controller wiring is always within one building, and where there is generally little need for telephone line

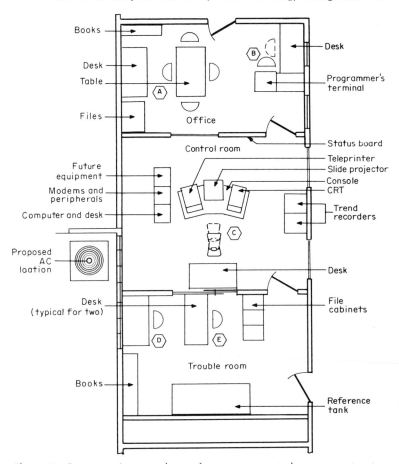

Fig. 4-14 Representative control room for an energy-control management system (ECMS): (A) ECMS engineer in charge; (B) system analyst; (C) ECMS dispatcher; (D) trouble board operator; (E) trouble board operator. (The Energy Group, Consultants, Los Angeles, California.)

interfacing, building location is not a significant cost consideration. Therefore, isolated buildings or small energy-using facilities without spare telephone lines lend themselves to local automatic control.

4-5-6 Typical Control-Point Combinations

The typical heating and cooling system control-point combinations in Table 4-1 can be grouped according to their purpose. The first six groups of points are used

TABLE 4-1 Representative Project Point List Summary

Purpose	Control Point	Type of Point			
		Binary	Analog	Start/Stop	Reset
Equipment shutdown	Fan motor	X		X	
	Water chiller	X		X	
	Chilled water pump	X		X	
	Condenser water pump	X		X	
	Cooling tower fan	X		X	
Outside-air shutoff	Outside-air dampers	X			X
	Exhaust fan	X		X	
Enthalpy control	Outside-air dry bulb temperature		X		
	Outside-air wet bulb temperature		X		
	Outside-air and return-air dampers	X			X
	Enthalpy controller				X
Enthalpy optimization	Outside-air dry bulb temperature		X		
	Outside-air wet bulb temperature		X		
	Return-air enthalpy		X		
	Outside-air and return-air dampers	X			X
Temperature reset	Hot deck temperature		X		X
	Cold deck temperature		X		X
	Chilled water supply temperature		X		X
	Chilled water return temperature		X		
	Condenser water supply temperature		X		X
	Condenser water return temperature		X		
Chiller load optimization	Chilled water supply temperature		X		X
	Chilled water return temperature		X		X
	Condenser water supply temperature		X		X
	Condenser water return temperature		X		X
	Chiller load ratio		X		X

TABLE 4-1 (Continued)

Purpose	Control Point	Type of Point			
		Binary	Analog	Start/Stop	Reset
Chiller maintenance	Motor status alarm	X			
	Chilled water pump status alarm	X			
	Condenser water pump status alarm	X			
	Chilled water supply temperature		X		
	Chilled water return temperature		X		
	Condenser water supply temperature		X		
	Condenser water return temperature		X		
	Differential oil pressure alarm	X			
	Cooler pressure alarm	X			
	Condenser pressure alarm	X			
	High motor temperature cutout alarm	X			
	Low refrigerant temperature cutout alarm	X			
	Starter overload alarm	X			
	Low water flow cutout alarm	X			
	Low oil pressure cutout alarm	X			
	Low condenser temperature alarm	X			
	Cooling tower status alarm	X			
	Excess chiller vibration alarm	X			
Boiler maintenance	Low water level alarm	X			
	Flame failure alarm	X			
	Inlet water temperature		X		
	Outlet water temperature		X		
	Boilers and hot water generators*	X		X	
	Boilers and Hot Water Generators Run Time*		X		
	Condensate receiver high-low level alarm*	X			
	Deaerator high-low level alarms*	X			
	Feedwater temperature*		X		

TABLE 4-1 (Continued)

Purpose	Control Point	Type of Point			
		Binary	Analog	Start/Stop	Reset
	Hot water flow*	X			
	Outlet Steam Pressure*		X		
	Pumps*	X		X	
	Safety Limits*	X			
	Stack Temperature*	X			
Air handling unit maintenance	Fan run time		X		
	Fan status alarm	X			
	Filter pressure drop alarm	X			
	Return-air temperature alarm	X			

*These control points are associated with boiler plant optimization and are not a part of any scheme.

SOURCE: Edward R. Berger and Edwin F. Coxe, "Automation and Centralization of Facilities Monitoring and Control Systems," ED 7601, U.S. Army Facilities Engineering Support Agency, Fort Belvoir, Va., May 1976.

primarily for conserving energy and are listed according to a particular energy conservation scheme. The last groups are aids in system maintenance and are listed by major equipment types.

4-5-7 Significance of System Selection

Although it might be desirable to procure control equipment whenever central control system expansion is made, this is not always possible. From control point to remote panels, the equipment can be procured competitively. However, each manufacturer employs a unique analog-to-digital data conversion, data acquisition, and data recovery methods. Currently, remote panels and remote point cards must be purchased from the original vendor. Future standardization may circumvent this limitation, Until then, it may be that per-point expansion costs are significantly higher than initial costs. Also, since the manufacturers' relative system costs vary drastically from application to application, different manufacturers will be economically attractive to different users.

4-5-8 General Range of Available Systems

The building automation system should pay for itself in a short time through energy and work force savings. The cost of automation systems, claims one manufacturer, varies with the number of points monitored in the system, the labor rates in different geographical areas, the building's physical design, and even the location of the mechanical equipment, including the location of field panels and the automation control center. Nonetheless, rough costs can be determined for different levels of automation.

The final decision on an automation system, says the same manufacturer, should offer the client the best balance between economics and capacity. One manufacturer has stated that the payoff for centralized building automation is about five years. With related energy-saving programs, the payoff period may be cut to three years.

In addition to savings in reduced energy consumption and reduced man-hours, a number of fringe benefits should be taken into consideration:

1. Longer equipment life and lower maintenance costs because of less operating time.
2. Reduced complaints from occupants.
3. More efficient record keeping and cost analysis.
4. Less probability of costly breakdowns.

Manufactured automation products can be divided into six basic marketing categories, called *levels of automation*. Each level represents the degree of sophistication of equipment and software. In general, the more points required, the higher the cost.

Level 1 Is represented by central consoles of the mini variety with a capacity of 100 points or less. Could include voice communication and set-point indication with reset. Is small enough to be mounted on a desk top. (Typical cost range: $3,000–$10,000.)

Level 2 Is typically represented by a free-standing cabinet. Handles up to 500 active system points. Optional equipment may include a trend recorder for analyzing system or individual point operation. Has start/stop programming. (Typical cost range: $15,000–$25,000.)

Level 3 Has 1000 active points plus the same optional equipment as for level 2. Could include optional 10,000-point capacity. (Typical cost range: $25,000–$40,000.)

Level 4 Is a desk-type command center with an alarm printer. Has optional 10,000-point capacity. Includes trend recorder and start/stop programmer. This level is the highest before computer memory is required. (Typical cost range: $40,000–$75,000.)

Level 5 Offers English-language alarm printout on a typewriter format. This level uses computer memory to perform data-logging functions. (Typical cost range: $90,000–$200,000.)

Level 6 Involves a *dedicated computer*. This means simply that the computer is completely free of other duties and is the exclusive slave of the automation command center. Complete building optimization is possible. (Typical cost range: $250,000 and above.)

Refer to Table 4-2 for a general description of features available for each level of sophistication.

4-6 AVAILABLE COMPUTER-MODELING OPTIONS

Effective computer modeling is based on developing each part of the computer model to produce thermodynamic identity to the actual system elements proposed.

There are a variety of proprietary and public computerized energy simulation programs on the market today. Each has its special features, but none seems able to do all things for all situations. Certain programs are useful in the systems simulation phase, others in load analysis, weather data utilization, equipment part-load performances, or economic analysis.

Most computer programs available for energy analysis in buildings present formidable obstacles to the first-time user. Frequently, it takes a great deal of time and effort to become familiar with just one of these programs. Because of this, consideration is often not given to similar computer programs which may be better for the intended purpose.[3]

There are various ways of comparing computer programs for energy analysis: several programs may be compared by one user, comparisons of the results on one program may be made by several users, or several users may make comparisons with several programs for the same problem. The results of these types of comparisons will frequently differ widely.

In a recent study,[3] a 20-story, 315,000-ft^2 office building in Washington, D.C., was analyzed by several computer programs. Actual weather data at Washington National Airport was used. Details were provided on all aspects of the building, which was rather conventional in design and typical of many built during the 1960s. The mechanical system consisted of a four-pipe fan coil system (which ran 24 hours a day) on the perimeter and a terminal reheat system on the interior. A single gas-fired hot water boiler and a single electric centrifugal chiller were the primary heating and cooling equipment.

Details on the part-load performance of all equipment, including boilers and chillers, were furnished by APEC (Automated Procedures for Engineering Consultants), a nonprofit group organized to develop computer programs. Information on the hourly load profiles of the various components of heating and cooling loads was also furnished. Month-by-month heating and cooling demands and consumption were recorded, as well as month-by-month gas and electric demands and consumption.

Figures 4-15 through 4-18 show the results of the study. Figures 4-15 and 4-16

TABLE 4-2 Building Automation Has a Number of Levels of Sophistication and Costs

Function of Automation Features		Level					
		1	2	3	4	5	6
Point capacity	100	x					
	500		x				
	1,000			x	x		
	10,000			o	o	x	x
Alarm indication		x	x	x	x	x	x
Status indication		x	x	x	x	x	x
Command	Start/stop/auto	x	x	x	x	x	x
	Summer/winter	x	x	x	x	x	x
	Increase/decrease	x	x	x	x	x	x
	Open/closed			x	x	x	x
	Temperature/humidity/checks	x	x	x	x	x	x
	Alarm summary		x	x	x	x	x
Indication	Set point	o	x	x	x	x	x
	Position	x	x	x	x	x	x
	Temperature/humidity/pressure	x	x	x	x	x	x
	Time			o	o	x	x	x
Voice communication			o	x	x	x	x	x
Trend recorders				o	o	o	x	x
System diagrams	Miniature graphic	x					
	Fullsize graphic		x				
	Rear screen projection			x	x	x	x
Programmed start/stop				o	o	o	x	x
Alarm printout	Coded				x		
	English language					x	x
Logging summaries	Point by point				x	x	x
	Selected systems					x	x
	In alarm points				x	x	x
	On/off status					x	x
Computer memory	Logging summaries				x	x	x
	Programmed start/stop					x	x
	Analog limit comparison					x	x
	Computerized maintenance						o
	Closed loop control						o
	Optimization						o

Code: x—standard, o—optional.

SOURCE: *Basics of Building Automation: Design Guidelines,* National Electrical Contractors Association, December 1973.

show the heating and cooling demands and consumptions on a month-by-month basis. The demands for heating and cooling indicate the highest peak hourly load imposed on the central system in each month. The consumptions for heating and cooling indicate the total monthly heating and cooling system loads, independent of the boilers and chillers used. Somewhat surprisingly, the cooling demands are in fairly close agreement, while the heating demands differ substantially. We might expect the opposite to be the case, since the cooling computations are much more complex. The probable reason for the wide variation in heating demands is the differing ways in which the "pick up" from the night setback condition was handled by the various computer programs. Heating and cooling energy consumptions in Figs. 4-15 and 4-16 differ considerably, probably because of how the various users interpreted the data in the computer programs and how the computer programs handled the heating and cooling loads, especially with the types of control systems specified by APEC.

Note that the summer consumption differences are generally greater than the winter consumption differences. This would appear to be because of the manner in

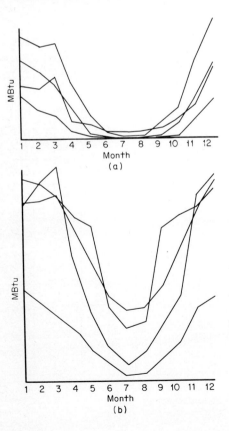

Fig. 4-15 Comparison of heating demand and consumption for various months of the year: (a) APEC symposium heating demand; (b) APEC symposium heating consumption. (Lawrence G. Spielvogel, "Comparisons of Energy Analysis Computer Programs," Automated Procedures for Engineering Consultants, Inc., HA-77-1, no. 1, Dayton, Ohio, 1977.)

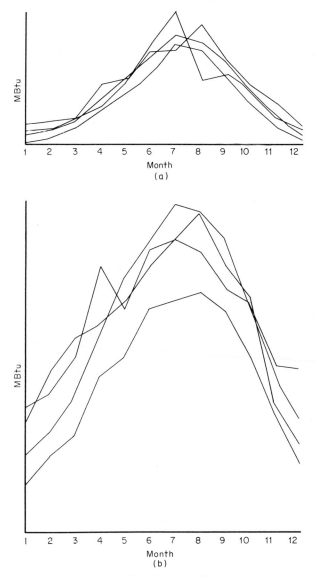

Fig. 4-16 Comparison of cooling demand and consumption for various months of the year: (a) APEC symposium cooling demand; (b) APEC symptosium cooling consumption. (Lawrence G. Spielvogel, "Comparison of Energy Analysis Computer Programs," Automated Procedures for Engineering Consultants, Inc., HA-77-1, no. 1, Dayton, Ohio, 1977.)

which the reheat systems were controlled and operated in the computer programs. Figure 4-17 shows the month-by-month gas consumption for all uses in the building. Here again, there is a fairly wide discrepancy among the programs, with the major differences occurring during the summer months. Figure 4-18 shows month-by-month electricity consumption. Except for one program, the results are in relatively close agreement.

The following conclusions were drawn from the above comparisons:

1. Results obtained by using several computer programs on the same building will range from very good agreement to no agreement at all. The degree of agreement depends on the interpretations made by the computer program user and on the ability of the computer programs to handle the building in question.

2. Several people using several programs on the same building will probably not get good agreement.

3. The same person using several programs on the same building may or may not get good agreement, depending upon the complexity of the building and its systems, and the ability of the computer programs to handle the specific conditions involved.

4. No evidence of any bias could be found in any of the programs utilized.

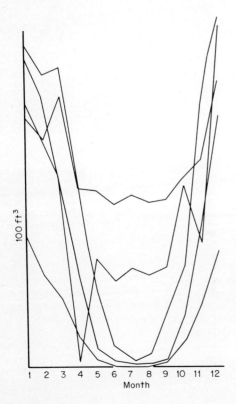

Fig. 4-17 Comparison of gas consumption for various months of the year (APEC symposium). (Lawrence G. Spielvogel, "Comparisons of Energy Analysis Computer Programs," Automated Procedures for Engineering Consultants, Inc., HA-77-1, no. 1, Dayton, Ohio, 1977.)

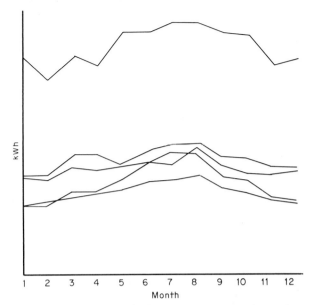

Fig. 4-18 Comparison of electric consumption for various months of the year (APEC symposium). (Lawrence G. Spielvogel, "Comparisons of Energy Analysis Computer Programs," Automated Procedures for Engineering Consultants, Inc., HA-77-1, no. 1, Dayton, Ohio, 1977).

Generally, environmental systems do not operate according to their original design; they require proper setup to reduce energy consumption. Installing load discriminators on single-zone, multizone, dual-duct, and terminal reheat systems will allow the automatic reset of temperatures in proportion to changing loads. Also, leakage of untreated outside air into the building often results in temperature and humidity control problems. Controlled pressurization reduces energy requirements necessary to handle such infiltration.

Some computer programs allow the introduction of part-load performance characteristics on some major pieces of mechanical equipment.[1,4] This feature can be used to select the best combination of equipment to meet the daily load demands. Many types of temperature controls are available to operate building systems. Most computer programs can simulate various types of HVAC systems and permit revised control-point settings to determine the optimum combination.

4-7 ALGORITHM DEVELOPMENT— ILLUSTRATIVE EXAMPLE

A number of algorithms can be employed in energy conservation programs. It may be instructive to look at one as an example of their general application.

The total electric power load of a major laboratory complex is illustrated in Fig. 4-19. This figure represents the Langley Research Center in Hampton, Virginia, throughout a particular day.

Before we can develop an appropriate algorithm, we must review a typical load profile to understand the differences in demand values for a typical utility customer under uncontrolled and under demand-limited situations. This is illustrated in Fig. 4-20. Assume that our demand-limiting system, through programming of appropriate algorithms, has shed (reduced) the demand during the peak period and redistributed it to some other time period. The daily energy consumption in kilowatthours remains the same; the areas under both curves should be identical. Yet the electrical costs become substantially lower for the demand-limiting case under conventional utility billing procedures.

Our system minicomputer can approximate the time period when the greatest electrical demand for our building will occur each day. One way it does this is by correlating average historical values of ambient and interior conditions with time periods of highest demand. Then adjustments are made between the corresponding historical time periods and current values of ambient and interior conditions. This

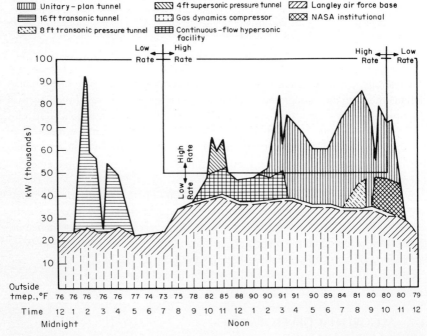

Fig. 4-19 Schematic of total electric power demand at Langley Research Center throughout a typical day. (Gerald P. Gaffney, "The Military Engineer," *Energy Conservation at NASA*, November/December 1975.)

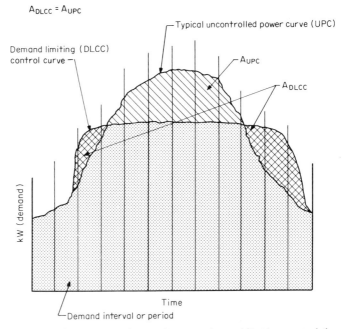

Fig. 4-20 Comparison of normal versus demand-limiting control for equal energy consumption. (Milton Meckler, "Energy Conservation Experience on Large Scale HVAC Systems," CH 77-12, delivered at ASHRAE semiannual meeting in Chicago, February 1977. Reprinted from *ASHRAE Transactions* by permission of the American Society of Heating, Refrigerating, and Air Conditioning Engineers, Inc.)

gives us the approximate probable time of greatest electrical demand. Such approximations should include seasonal effects and variations in annual heating and cooling loads.

Once we have developed a rationale for identifying the critical daily demand period, a variety of operating strategies become possible. During critical periods, we can:

1. Precool the chilled water (storage) system and unload the chiller using available thermal capacity to maintain space conditions during this time period.
2. Deenergize noncritical loads, such as tower or exhaust fans, pumps, domestic electric water heaters, and outdoor lighting.
3. Stage operation of electric heaters into smaller sections, rotate and automatically cycle each on some basis, such as 30 seconds on/30 seconds off.
4. Provide for automatic damper actuation of air-handling units from minimum outside to a 100 percent return-air position, (provided atmospheric cooling or critical ventilation needs are not concurrently demanded).

By way of our algorithm, we can establish override situations for each of the above ECMs. These can be controlled by predetermined space temperature and humidity limits so that, in extreme conditions, demand lighting, for example, can be sacrificed in lieu of critical environments.

See Fig. 4-21 for a schematic representation of our proposed demand-limiting algorithm. We must recognize that each building has unique requirements which may entail some variations to the following proposed steps:

1. Step A represents the point of minicomputer entry under normal conditions, programs normal housekeeping functions, etc.

2. Step B queries system to determine if predetermined control point will be exceeded in present demand period. Before responding, the minicomputer must extrapolate trends in the instantaneous power demand(s) and anticipate any additional electrical loads that may be called for by other building system programs and/or HVAC set-point adjustments during the balance of the "sampled" time interval.

3. If the answer in step B is affirmative, the minicomputer automatically advances to step C, where some load-shed decision is made based on information available to step F. It is assumed that there are sufficient electrical loads of lowest priority capable of being reduced to keep the anticipated electrical demand below the predetermined control limit.

4. The minicomputer now enters step D, where the electrical loads selected in item 3 above are shed.

5. After shedding, the program enters a delay mode (step E), to allow sufficient time to accomplish the designated load sheds before returning to step B.

6. Should the answer to step B be negative (no), the minicomputer must interrogate step G to determine if any electrical loads are presently in a shedding mode. If affirmative, the program automatically advances to step H to establish that the need for this shedding has passed. This decision is a function of the differences in actual and current demand values as determined from computations made in step B. If this difference is below some predetermined value, the electrical load sheds under step G cannot be reinstated, and the minicomputer automatically exits the program under step I. If, however, this difference is above the predetermined value, the minicomputer reinstates the most recently shed loads under step J and then automatically proceeds back to step E.

7. If a negative result had been obtained under step G, the minicomputer would have determined that the current trend would not cause the demand limit to be exceeded (in this time interval) and that no electrical loads are currently shed. Where this is the case, the minicomputer advances to step K and computes the anticipated electrical load for the next several hours, drawing upon data stored in steps $O, P,$ and Q. The resulting demand profile is then stored in step L.

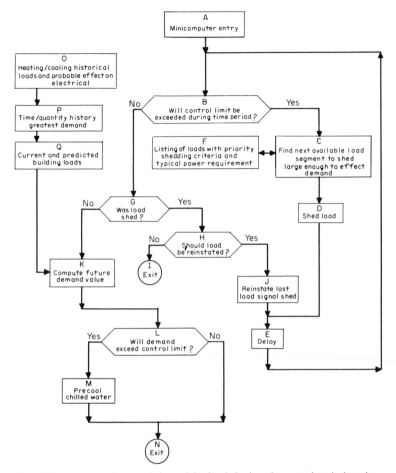

Fig. 4-21 Suggested program logical for load shed cycle control including thermal (chilled) water diversity. (Milton Meckler, "Energy Conservation Experience on Large Scale HVAC Systems." CH 77-12, delivered at ASHRAE semiannual meeting in Chicago, February 1977. Reprinted from *ASHRAE Transactions* by permission of the American Society of Heating, Refrigerating, and Air Conditioning Engineers, Inc.)

8. The above profile (step *L*) is interrogated to determine if the demand limit is likely to be exceeded in the next several hours. If an affirmative result is reached, the program automatically advances to step *M*, where precooling of the chilled water system is initiated. The minicomputer then exits the program under step *N*. If a negative result is arrived at under step *L*, the minicomputer exits the program at step *N* directly.

4-8 COMPUTERIZED ENERGY MANAGEMENT TRENDS

Building energy management requires differing levels of computer technology, tailored by the consultant to specific owner needs. Selecting the best configuration is often a complex task. Design practices are also changing rapidly. New types of hardware are radically expanding the opportunities for computerized energy management. Owners and users, confused by conflicting vendor promises, often have little actual knowledge of their real energy management needs, or of what the consultant can do to maximize their energy conservation.[5]

Traditionally, centralized building control systems have been provided by a half-dozen companies active in the HVAC control field. As security, fire detection, and allied requirements become increasingly important to the building owner, these functions were integrated into central control systems.

Now, however, several companies with electronics, aerospace, and military equipment experience have entered the field. Their expertise, as it relates to building automation, resides principally in the fields of digital transmission, multiplexing, and electronic instrumentation.

Additionally, consulting firms are being established to offer services in the professional management of building operations. Their goal is to provide owners with realistic operating cost programs. These can include the operation of building systems, repairs (remedial and projected), maintenance and scheduling, and instruction for operating engineers.

SUMMARY

Because buildings and their internal systems have grown larger and more complex, computers, microprocessors, and automated energy management systems have assumed an important role.

When selecting computer hardware and software, early planning is advised; control monitoring flexibility and expanding capacity capability should be considered at the onset.

A new generation of computer control systems is now becoming available. In addition to the normal functions of other building automatic systems, they are able to monitor and measure the distribution of all energy types, identify the sources of inefficient energy consumption, develop algorithms for controlling and optimizing this consumption, and modify these algorithms when necessary.

There are three levels of control possible in a building automatic control system: local control, central manual control, and central automatic control.

Some useful computer programs commercially available as programmable software or hard-wired from major control manufacturers include enthalpy optimization programs, variable-start-time selection programs, chilled water optimization programs, electrical demand load-leveling programs, electrical demand load-shedding programs, load profile programs, and power-factor correction programs.

Four major factors to be considered when selecting computerized control systems are building size, usage patterns, heating and cooling systems, and relative building location.

The cost of computer systems depends on factors such as the number of points monitored within the system, local labor rates, the building's physical design, and the location of mechanical equipment.

Selecting the proper computer hardware and the appropriate software options can be a difficult and confusing task; the field is rapidly expanding, and there are many conflicting vendor claims and promises. Increasingly, building owners are seeking the services of consulting firms skilled in the field of automatic building control systems.

NOTES

1. M. Meckler, "Energy Conservation Experience on Large Scale HVAC Systems," CH-77-12, delivered at ASHRAE semiannual meeting, Chicago, February 1977.
2. Edward R. Berger and Edwin F. Coxe, "Automation and Centralization of Facilities Monitoring and Control Systems," ED 7601, U.S. Army Facilities Engineering Support Agency, Fort Belvoir, Va., May 1976.
3. L. G. Spielvogel, "Comparisons of Energy Analysis Computer Programs," Automated Procedures for Engineering Consultants, Inc., HA-77-1, no. 1, Dayton, Ohio, 1977.
4. *Basics of Building Automation: Design Guidelines*, chap. 4, p. 74, National Electrical Contractors Association, December 1973.
5. Gregory P. LaRocca, "Computerized Energy Management," *Electr. Consultant Eng. Electr. Energy Sys.* (October 1975).

SELECTED BIBLIOGRAPHY

NOTE: *Reports without listed authors are available through the U.S. Government Printing Office or the National Technical Information Service.*

"Automation and Centralization of Facilities Monitoring and Control Systems," U.S. Department of Commerce No. AD-A026 693/2WAY, National Technical Information Service.

Bizarro, Leonard A.: "Networking Computers for Process Control," *Chem. Eng.* (December 1976).

"A Comprehensive Computer-Aided Building Design System," Massachusetts Institute of Technology, National Technical Information Service, PB-212 615, June 1972.

"A Directory of Computer Software Applications, Energy," U.S. Department of Commerce No. PB-264 200/7, National Technical Information Service.

"Evaluation of Computerized Energy Programs for the Calculation of Energy Use in New and

Existing Buildings," U.S. Department of Commerce No. PB-272 337/7, National Technical Information Service.

Evans, L., and R. Jones: "Successful Applications of Minicomputers in Manufacturing," Technical Paper MS76-266, Society of Manufacturing Engineers, Dearborn, Mich., 1976.

Gaffney, Gerald P.: "Energy Conservation at NASA," *Mil. Eng.* (November/December 1975).

Hittle, Douglas C., and Dale L. Herron: "Simulation of the Performance of Multizone and Variable Volume HVAC Systems in Four Geographical Locations," *ASHRAE Trans.*, **83,** pt. 1 (1977).

Maugh II, Thomas H.: "Microprocessors: More Instruments Are Becoming Smart," *Science,* **199** (March 1978).

"Methodology of Technology Analysis with Application of Energy Assessment," Brookhaven National Laboratories, Upton, N.Y., U.S. Department of Commerce No. BNL-20083, National Technical Information Service.

"Modeling and Simulation in Power Plant Processes," U.S. Department of Commerce No. KFK-PDV-90, National Technical Information Service.

Mueller, Dr. George E.: "Whiter the Large Computer," *Mini-Micro Sys.* (January 1977).

NBSLD (National Bureau of Standards Load Determination), "The Computer Program for Heating and Cooling Loads in Buildings," U.S. Department of Commerce No. PB-267 169/1, National Technical Information Service.

Ogdin, Carol A.: "Fundamentals of Microcomputer Systems," *Mini-Micro Sys.* (November/December 1977).

Peterson, Jeffrey N., Chau-Chyun Chen, and Lawrence B. Evans: "Computer Programs for Chemical Engineers," *Chem. Eng.,* pt. 1 (June 1978).

———: "Computer Programs for Chemical Engineers," *Chem. Eng.,* pt. 2 (July 1978).

"A Review of Computer Software Applicable to the MIUS Program," U.S. Department of Commerce No. PB-273 175/OWE, National Technical Information Service.

Shaffer, E. W.: "Computerized Approach to Energy Conservation," Technical Paper EM76-107, Society of Manufacturing Engineers, Dearborn, Mich., 1976.

Shapiro, Richard B.: "Mastering the Micro," *Mini-Micro Sys.* (November/December 1977).

"Simplified Algorithm to Solve Geometric Programming Problems Using Fortran," U.S. Department of Commerce No. COO/2895-6, National Technical Information Service.

Energy Conservation and Load Management

Energy conservation is the most important, and the most ignored facet of energy policy.

Dennis Hays

The Morgan Guaranty Trust Company has warned that the safety margin (the amount of electric generating capacity above peak level demands) could drop to as low as 14.5 percent by 1986 from a present level of approximately 30 percent. The supply side of the electrical energy equation needs urgent attention. Approximately 12 years are required to bring a new nuclear plant on-line (as opposed to 10 years for a comparable fossil-fueled plant). Even with a projected slower growth of demand, electric power shortages by 1986 could disrupt industry, hamper farms, aggravate unemployment, impair GNP growth, and bring about a reduced standard of living.

Until the recent past, the electric utility industry had been a declining-cost industry. The cost of producing an additional kilowatthour of electricity had been going down steadily. The reasons for this were technological. As electricity generating plants were built in larger sizes with continuing technological improvements, their greater efficiencies were translated directly into lower prices.

Now, however, buildings and industrial facilities share a common concern, particularly if they are located in one of the energy problem areas of the country such as the Northeastern sector. Problems are particularly worrisome in metropolitan New York and Boston, where both reduced energy availability and high cost prevail. Various utilities are engaged in ongoing studies and experimental and demonstration programs aimed at reducing peak electrical demand at minimum total cost. A number of novel control devices for the direct management of customers' loads have proven efficient and cost-effective.

Various indirect utility programs involving rate incentives and disincentives are gaining broad acceptance among the public regulatory agencies despite insufficient data pertaining to their effectiveness or fairness.

5-1 MANAGING CUSTOMER LOAD PROFILES

Electrical demand (the rate of power consumption) is monitored by utility companies to establish the maximum rate of energy usage by their customers. The utilities reason that, since they must provide generators, transformer, etc., sized to accommodate the maximum power rate, the maximum rate can be used to "equitably distribute" applicable capital costs.

This means that the consumer must redistribute (shed) designated portions of the electrical load during the maximum demand period, without reducing the total kilowatthours consumed, to save on utility bills.

In an effort to cut the electric load demand of commercial and industrial users, some electric utilities are urging load leveling and/or shifting and greater energy efficiency. These efforts are aimed at reduced reliance on our nation's (often less efficient) fossil-fueled peaking units. They should also reduce the need for adding costly generation facilities by cutting coincident loads at or near the system peak demand (see Fig. 5-1).

To understand load management problems as viewed by the electric utility, we must first understand load management in terms of progressive utility growth in a given service area. Some representative daily load-duration curves are illustrated in Figs. 5-1 through 5-4. Figure 5-1 represents the profile of a typical unmanaged load. It is characterized by a steep slope and a low ratio of overall output generally concentrated to result in a short duration in peak demand. This requires that standby utility peak generating capacity be activated. The utility still generates the same total energy output by distributing it more evenly in the manner shown in Fig. 5-2. Notice that the initial program efforts have resulted in an average demand which remains unchanged while representing a higher ratio of the reduced peak, resulting in a higher utility load factor. Although the fixed cost of capacity per kilowatthour remains

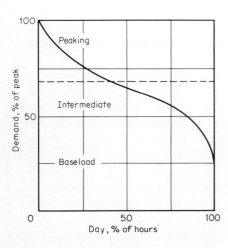

Fig. 5-1 Short-duration, peak-demand daily load-duration curve. Load management and the progressive effects of system growth are shown in simplified form by this and the three following daily load-duration curves. Unmanaged load has an appreciable slope, a small proportion of overall output being concentrated to create a relatively short period of peak demand and a requirement for peak generating capacity to fill it. (Ken Owens, "Demand Control Is a New Game Plan for Utilities," *Electrical Consultant*, September/October 1977, p. 32.)

Fig. 5-2 Improved-distribution daily load-duration curve. Managed load involves the same total energy output, but the load is distributed more evenly. The unchanged average demand is a higher proportion of the reduced peak; in other words, the load factor is higher. Capacity cost per kWh is unchanged, but fuel and operating costs are down because there is less use of peaking units. Reliability is up because of greater idle capacity. (Ken Owens, "Demand Control Is a New Game Plan for Utilities," *Electrical Consultant*, September/October 1977, p. 32.)

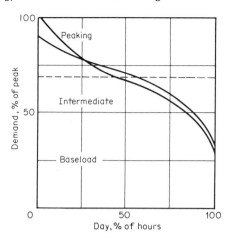

constant, the operating costs associated with fuel, operations, and maintenance are reduced, since there is a lesser requirement for employing peaking units.

A further benefit is that generally system reliability is improved, since there is a greater margin of idle capacity. In Fig. 5-3, we assume some time has elapsed since the program implementation efforts represented in Fig. 5-2. Figure 5-3 reflects a managed load with the same load factor but with system growth. Notice that the peak demand falls below the earlier (or original) value (see Fig. 5-1). Therefore, system reliability also remains unchanged. Although capacity cost is reduced, the increased

Fig. 5-3 Managed-growth daily load-duration curve. Managed load with growth, after time has passed, has the same load factor as the managed load, but the peak demand is now just below the original unmanaged value, and reliability is now again the same as for the unmanaged load. Capacity cost is less because an unchanged capacity mix is producing more energy at the high load factor. But peaking units are running more than in the unmanaged case, and high fuel and operating costs offset much of the capital cost saving. (Ken Owens, "Demand Control Is a New Game Plan for Utilities," *Electrical Consultant*, September/October 1977, p. 32.)

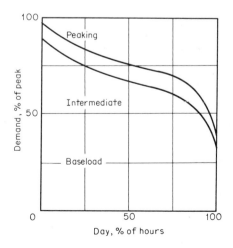

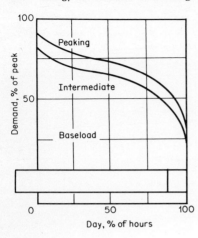

Fig. 5-4 Managed-load/capacity daily load-duration curve. Managed load with new capacity, after still more time, accommodates a still higher overall load. But the capacity mix is now being optimized for the new load factor by adding baseload units, which take over from less efficient intermediate and peaking units to cut fuel and operating costs again. (Ken Owens, "Demand Control Is a New Game Plan for Utilities," *Electrical Consultant*, September/October 1977, p. 33.)

running time of system peaking units tends to erode those savings due to associated higher fuel and operating costs.

This problem is alleviated somewhat by the addition of new capacity as illustrated in Fig. 5-4, which must accommodate a higher load than that illustrated in Fig. 5-3. A basic difference between these figures is that the capacity mix is optimized for the new load factor by selective addition of base-load units, which now replace the less efficient intermediate and peak requirements. This reduces fuel and operating costs, thereby breaking the cost spiral through an effective, planned utility load (growth) management program.

5-2 ENERGY STORAGE SYSTEMS

For almost all renewable energy sources, supplementary energy must be available when conditions (wind, solar, etc.) are insufficient for ample power generation. This requires installation of an energy storage system capable of storing excess energy when generation exceeds demand and retrieving it under reversed conditions. Refer to Table 5-1 for a comparison of some commonly employed or suggested energy storage systems.

Various methods of using vapor compression refrigeration chillers operating simultaneously as heat pumps have been analyzed and compared with conventional central heating and cooling plants.[1-3] In such systems, both heating (split-condenser or double-bundle) and cooling chiller capacity are utilized by building HVAC systems and terminals. Unitary water-to-air heat pump systems with thermal energy storage capacity can offer the same features on the heating cycle, often at less first cost and with a modularity that reduces auxiliary energy usage while maintaining the same

TABLE 5-1 Characteristics of Energy Storage Technology

Technology	Typical Economic Module, MWe	Characteristics		Remarks
		Earliest Commercial Availability	Storage Efficiency, %	
Batteries	1	1975–1982	70–80	Proven technology
Flywheels	1	1985	70–90	
Hydrogen/fuel cells	1	1985	40–60	Storage options for hydrogen
Compressed-air:				
Adiabatic	10 (30 MWh)	1982	70–80	High-grade thermal
Isothermal	10	1975	NA	Required fuel
Pumped hydro	100 (?)	1975	70–75	Special situations
Superconducting magnets	500	1995	90	

SOURCE: M. Zlotnick, *Proceedings of the Second Workshop on Wind Energy Conversion Systems*, Washington, June 9–11, 1975, NSF-RA-N-75-050.

diversity advantages as the chiller on cooling cycle.[4-6] A number of electric utilities[7] have been urging their customers to consider thermal energy storage (TES) in large industrial and commercial installations. Many office buildings, dormitories, motels, and hospitals have converted to all-electric or merely added TES after learning that the controllable load in most buildings can be served at less cost if a properly designed off-peak storage system is used.

The application of low-temperature TES reduces the purchased energy consumption further, makes it possible to use a smaller refrigeration machine, and levels the peak building cooling load in the manner illustrated in Fig. 5-5. Load leveling reduces wasteful transients as well as utility load peaking, which is costly and results in power shortages, particularly during peak hot weather periods.

TES can also be utilized effectively in solar power generation facilities.[8,9] A thermal storage vessel containing iron with an assumed 30 percent voids, when heated to 1500°F (816°C), has a resultant heat capacity of 20,000 Btu/ft^3 (745,355 kJ/m^3). It would be capable of generating steam at 1100°F (600°C) for efficient steam generation systems. With a 33 percent power plant efficiency level, a high-temperature thermal storage tank volume of 51,000 ft^3 (1444 m^3) could produce 100 MWh of electrical energy. If we employ instead a lithium hybrid which changes from solid to liquid when heated to 1200–1300°F (649–704°C), the storage volume required to produce the same 100 MWh would take roughly 20 percent as much storage volume at 11,000 ft^3 (311 m^3). TES can also be used by utility customers on a decoupled basis to produce power at periods of low demand. It is then stored to help trim electrical use during periods of utility peak demand. Practical TES is now well

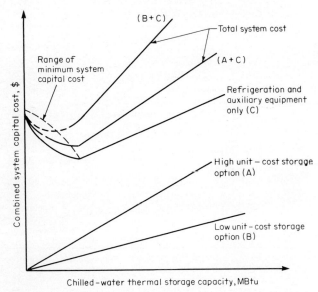

Fig. 5-5 Relationship of chilled water storage to total capital requirements of associated refrigeration plants using a combination of refrigeration cycle and low-temperature (chilled water) thermal energy storage.

within the state of the art. Costs associated with TES can generally be amortized within a reasonable period of time for those customers served from a utility which offers low enough off-peak or time-of-day rates.

TES systems work in a variety of different ways to accomplish the same result. Figure 5-6 illustrates a hypothetical utility load curve for a peak summer weekday. TES allows the customer to refrain from using electricity deferred from "peaking" generation, covering the time period of roughly 0900 through 2300 hours. Aside from reduced billings, further conservation benefit through advantageous fuel substitution is also accomplished.

Here is a brief description of representative TES options:

1. Nonpressurized chilled and hot water storage, requiring interconnection with both a chiller and heat source.

2. Nonpressurized hot water storage vented to the atmosphere with utilized thermal energy supplied from solar, heat pump, or electric resistance means.

3. Pressured water storage in a closed vessel, served from heat sources similar to those discussed in option 2 but incapable of higher storage-temperature levels.

4. High-temperature heat storage systems operating at temperatures up to 1400°F (760°C) and heated by electric resistance means.

5. Static ice storage, utilizing heat or fusion of water in addition to available sensible heat from environment.

6. Same as above, except in combination with hot and chilled water storage and employing a single tank.

7. Annual cycle energy system (ACES) which utilizes water and ice to store heat during summer months and dechilled water during winter months by means of a large storage container employed over a longer period of time to accomplish the desired seasonal cycle effect.

8. In-ground (heated) storage employing electric resistance elements, hot water, or hot water transfer via heat pipe either to change the ground temperature below or adjacent to a building, respectively. Provides direct radiant heat or heat source for winter months or serves as a heat sink during summer months.

The use of TES is expected to accelerate as electric rates rise and as off-peak or time-of-day pricing becomes more widespread.

It has been suggested that flywheels operating in low-pressure inert atmospheres (to reduce drag losses) can store 10 to 20 × 10^3 kwh at a rotational speed of 3500 r/min and an estimated cost of $110/kW. This is competitive with current emergency power generation systems. More importantly, such flywheel systems operate at an output-to-input energy efficiency of 90 to 95 percent, as compared to the 50 to 70 percent output-to-input efficiencies of comparable pumped storage systems (which also are approximately 2.5 times more costly to build).

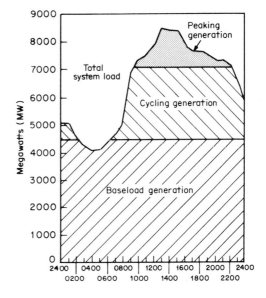

Fig. 5-6 Distribution of generation capacity by type for daily load profile case. (Jim McCallum, "TES Plus Off-Peak Can Save Energy and Lower Cost," *Air Conditioning & Refrigeration Business,* October 1977.)

5-3 UTILITY RATE IMPACTS

While changes in the cost structure of the utility industry have been taking place, there has also been widespread concern about the impact of continued growth in electricity consumption, the quality of our environment, and our dependence on foreign fuel oil.

All this has focused attention on methods to reduce past patterns of growth in energy consumption. These methods include changes in the rate structure. It is now widely believed that the traditional declining block rate structure, which charges less for the one-thousandth kilowatthour than for the one-hundredth, leads to continued growth in consumption.

The choice of the method for determining costs is important because it serves as a signal to buyers and sellers in the marketplace. The prices at which customers can buy electricity at any time communicate to them the costs of producing energy at that time. Correspondingly, if customers are willing to pay an amount equal to or above the current selling price, they are signaling the seller to supply more electricity at that time. In theory, this system serves two important economic functions:

1. It promotes efficient use of society's scarce capital and energy resources.
2. It distributes the costs most appropriately among customers.

Prices set on the basis of marginal cost promote efficient energy use by signaling the customer what it costs to consume a little more energy—and conversely, what the utility saves in system cost if the customer consumes a little less energy. In this way, prices assure that no customer consumes electricity which is worth less to the customer to consume than it is worth to society to produce. When electricity is underpriced or overpriced, the pricing structure either encourages waste or it unnecessarily deprives some customers of energy use or both. When rates are based on average cost today, energy is necessarily underpriced at the very time of the day it is more expensive to produce (at the time of the system's peak demand), and it is overpriced at the very time it is least expensive to produce (at times of relatively low system demand).

Peak-period pricing provides the most attractive means of overcoming the deficiencies in a utility system's rate structure. It is most consistent with the basic principles outlined above. Both time-of-day rates and seasonal rates are being considered, but, given the greater cost differentials between various periods of the day than between the seasons, the time-of-day rates appear to be more important.

In spite of a generally limited peaking profile, utilities around the country appear to be offering meaningful time-of-day rates to industrial customers. According to Edison Electric Institute (EEI) studies published recently, mandated time-of-day rates are in effect for some 19 states. Experiments are being conducted in 24 states. Only a few generally nonindustrialized states have no plans for time-of-day rates.

The details of a time-of-day tariff will depend on the specific circumstances of each electric utility. Circumstances may change as patterns of energy use and the equipment available for generation change. In the Los Angeles Department of Water and Power system, the pattern of costs is such that a single peak-period rate and a single off-peak rate have been recommended as an initial tariff. The following rationale has been advised: The on-peak charge for energy should reflect the marginal cost of fuel to produce energy during the peak period of the day, plus a more than proportionate share of the actual generation and transmission costs. The off-peak charge should reflect the fuel cost of operating the most efficient configuration of generating facilities. A small charge, based on individual maximum kilowatt demand, may also be made. Its purpose is to recover customer-specific marginal capital costs of the transformer and distribution system immediately connected to the maximum kilowatt demand that the customer is ever expected to impose on the system and any special metering equipment associated with the customer's services.

Those customers who impose above-average costs on the system will see those costs reflected in their bills. Customers imposing below-average costs on the system, by consuming electricity at times other than the peak demand period, will see those savings reflected in their bills.

The most precise representation of system costs will be achieved if most of the cost difference is reflected in the kilowatthour charge. By minimizing the importance of the kilowatt charge, the utility conveys to the customer the principal message of system costs—it is not the momentary kilowatt demand placed on the system by the individual customer that is important but rather the number of expensive kilowatt-hours consumed by the customer during the system peak period. Figure 5-7 illustrates this distinction by showing the electrical demand over a 24-hour period for three customers: customers A, B, and C.

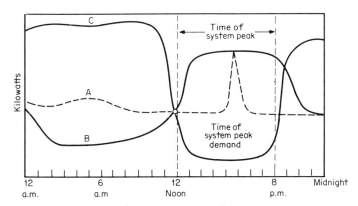

Fig. 5-7 Representative variations for time-of-day rates. (Milton Meckler, "Applying Cost Effective Solar Assisted Heat Pump Systems Now," *Proceedings of the First National Conference on Technology for Energy Conservation*, June 1977, Library of Congress cat. no. 77-91045.)

All three consume the same total number of kilowatthours over the entire day and have the same peak kilowatt demand. The *system* peak demand occurs from 12 noon until about 6 p.m. Customer A uses electricity at a very constant rate except for a brief period each afternoon. Customer B's pattern of use varies considerably; more than half of B's consumption occurs during the period of system peak.

Under many current rate structures, both customers pay identical amounts for their electricity use; however, customer B imposes significantly greater costs on the system than customer A. A tariff based on a time-differentiated kilowatthour charge would bill customer B a greater amount than customer A. This would more accurately reflect the greater costs B imposes on the system.

Customer C has the same kilowatthour energy use and kilowatt demand as customers A and B, but only a small fraction of C's total energy is consumed during the system peak period. Almost all of C's energy, including individual peak amount, is consumed outside the system peak period. This customer is the least costly to serve, yet many current departments of water and power rates charge customer C exactly the same amount as customers A and B.

5-4 ELECTRICAL DEMAND AND ALTERNATE ENERGY SOURCES

It is often convenient to justify a solar heating and cooling system on the basis that its incremental additional capital cost is offset by the annual cost savings of the energy resources it displaces over the system's lifetime.[10,11] The higher the cost of the displaced energy source, the more acceptable the solar alternative becomes. If the energy cost is very low, there will not be enough cost savings to justify the system. The greater the difference between cost of supply at peak and off-peak operating periods, or the larger the time-of-day variation in the cost of supply, the easier it is to justify the incremental capital cost of a load management system employing thermal storage and associated controls. Where the time-of-day variation in the cost of supply is small, the potential for chilled water, load management cost savings may still justify additional equipment cost, since the chilled water serves to reduce the installed capacity of the various heat pump units by providing more diversity.

The question is whether the reduced utility cost is enough to offset the added capital cost of thermal energy, additional piping, and appurtenances. In this sense, we can employ chilled water as a load-shedding strategy.[12]

DESIGN CONSIDERATIONS FOR LOAD MANAGEMENT

HVAC and other mechanical systems should be designed and selected with the realities of load management firmly in mind. User requirements, utility peaking periods, and equipment characteristics must be carefully balanced. Energy storage systems should be considered for accumulating energy during off-peak hours. The designer may also wish to consider alternate energy sources as an energy augmentation strategy.

SUMMARY

Electrical demand is growing, and the nation's generating capacity is failing to keep pace. Costs are rising and various rate incentives and disincentives have been gaining broad acceptance. Specifically, the practice of load shedding, where the consumer redistributes designated portions of the electrical load during the maximum demand period, is being urged by many utilities.

Energy storage systems, such as thermal energy storage (TES) systems, can be employed to store energy during off-peak periods for use during the maximum demand period.

Utilities in many parts of the country are changing their rate-price structures to encourage conservation and to reduce the need to add new generating capacity. Mandatory time-of-day rates have been put into effect in many states. Other states are experimenting with similar arrangements.

Finally, alternate energy sources, such as solar-powered heating and cooling systems, should be considered when justified by large variations in the time-of-day rates.

NOTES

1. M. Meckler, "Heat Reclamation Strategies," *Buildings* (November 1976).
2. Michael Baker, Jr., "Design Guidelines for Energy Conserving Systems," U.S. Department of Commerce Report No. PB-268-989, March 1977.
3. "Solar Energy Heat Pump Systems for Heating and Cooling Buildings," ERDA Doc. C00-2560-1, Con. 7506130, The Pennsylvania State University, National Technical Information Service.
4. M. Meckler, "Cost Effective Solar Augmented Heat Pump/Power Building Systems," 23rd Annual Technical Meeting and Equipment Exposition of the Institute of Environmental Sciences, Session A7: Solar Energy Utilization, Institute of Environmental Sciences Proceedings, Apr. 27, 1977.
5. M. Meckler, "Applying Cost Effective Solar Assisted Heat Pump Systems Now," Proceedings of 1977 National Conference on Technology for Energy Conservation, June 8–10, 1977, Washington, Library of Congress No. 77-91045.
6. M. Meckler, "Simulations of Solar Assisted Multiple Zone Water Source Heat Pump System with Diversity Cooling Loop," vol. 2.1, Proceedings 1978 Annual Meeting of the American Section of the International Solar Energy Society, Aug. 28–31, 1978, Denver, Col.
7. Jim McCallum, "TES Plus Off-Peak Rate Can Save Energy and Lower Cost," *Air Cond. Refrig. Bus.* (October 1977).

8. M. Meckler, "Solar Energy and Large Building HVAC Systems: Are They Compatible?", *ASHRAE J.*, pp. 43–50 (November 1977).

9. M. Meckler, "Solar Energy Selective Power HVAC Systems for Large Building Applications," Proceedings of the 1978 National Conference on Technology for Energy Conservation, Jan. 24–27, 1978, Library of Congress No. 78-56277.

10. Craig H. Peterson, "The Market Capture Potential of Single vs. Multistructure Solar Energy Space Conditioning Systems: 1975–2010," prepared for the National Science Foundation Research Applied to National Needs, January 1976.

11. Proceedings of the First Semiannual SPRI Solar Program Review Meeting and Workshop Held in San Diego, Calif., On Mar. 8–12, 1976, vol. 1, "Solar Heating and Cooling of Buildings," U.S. Department of Commerce no. PB-260 594, National Technical Information Service.

12. M. Meckler, "Energy Conservation Experience on Large Scale HVAC Systems," *ASHRAE Trans.*, **83,** pt. 1 (1977).

On-Site Energy Systems

Neither conservation, exotic power, nor standard plant designs seem likely to forestall the predicted near term shortages of generating capacity.

R. C. Rittenhouse

An on-site energy system (OSES) may be defined as an arrangement in which all (the total energy option) or some (selective power option) of the requirements of an energy-using facility are integrated to maximize the utilization of primary fuel sources.[1] In a more general sense the terms *OSES, cogeneration, on-site generation,* and *combined-cycle plants* have come to be used interchangeably to describe the production of useful heat and electricity at the same location. As indicated above, such on-site energy plants may be of the total energy or the selective power type. Selective power applications generally result from a surplus of waste heat available at the maximum electrical load requirement. The thermal loads, only, are balanced with the corresponding electrical load to reduce purchased power.

Methods employed in the design of OSES' include the recovery of waste heat from heat engines which generate electricity or drive mechanical equipment, the application of energy-using equipment that gives a balance between mechanical-energy waste heat, and the integration of other utilities, such as water and waste treatment, into the energy system.

Employing heat normally wasted, OSES' designed to operate at high thermal efficiencies can achieve substantial savings in operating costs. The average electric utility generating plant normally converts its fuel at an efficiency of from 25 to 35 percent.[2] In contrast, the average OSES can, if combined with the mechanical and electrical systems, utilize roughly 65 to 70 percent of the chemical energy in the fuel consumed. This depends somewhat on energy loads, profiles, and balances.

Fortunately, many industrial, commercial, and institutional complexes now have processes capable of utilizing waste heat. It is also possible to install energy-consuming equipment which relies on heat instead of power: absorption cooling instead of mechanical cooling, for example. Such an arrangement can simultaneously reduce overall power requirements and increase the application of waste heat. A further advantage of on-site generation can be improvement in power quality; fewer outages and more precise control of electrical characteristics, often needed for electronic data processing, are possible.

While on-site generation with heat recovery can lead to substantial savings, these reductions can be realized only under proper conditions. A thorough analysis is essential to determine whether or not such conditions exist for a specific facility.[3,4]

6-1 HISTORICAL PERSPECTIVE

Most of the electricity generated in the United States originates in large central station plants owned and operated by the electric utilities.

OSES' have been designed for diesel or steam plant prime units and for combinations of the several prime-unit types.[5] Although an OSES appears very attractive because of its efficiency and flexibility, OSES' have not been expanding in the United States in recent years. To cope with normal maintenance or unscheduled outages, provisions for utilizing power from a local central power station may be required. The cost of having auxiliary power available for outages tends to offset some of the obvious advantages of the OSES. The light fuels used by OSES' are relatively expensive, and operation and maintenance of the gas turbine or diesel units require specialized skills.

The number of conventionally fueled on-site generation plants has been steadily declining, while the average size of utility generating plants has increased. In 1930, on-site generating facilities represented 30 percent of all U.S. generating capacity expressed in kilowatthours generated. This declined to only 4.2 percent in 1973. Yet during the short period from 1965 through 1974, the percentage of electric utility plants in excess of 500-MW capacity increased from 40 to 56 percent. This reflected a conviction that economy of scale would permit more efficient plant operation and a lower cost to customers. The trend toward centralization was founded on the following key factors:

1. The belief that the capital cost per unit of installed capacity of utility generation and associated auxiliary equipment decreases as plant size increases.
2. The belief that the larger the plant, the lower its operating cost per unit of output.
3. Siting problems brought on by environmental constraints suggested removing the larger, more efficient, plants from population centers.
4. Larger plants favored the use of coal.
5. Rate benefits to customers would be reflected in lower average rates per kilowatthour as utility sales expanded.

A growing conviction, reinforced by major promotional and sales efforts, was that the headaches of electricity production (such as noise and air pollution, utility opposition, failure, and inexpert maintenance) were better left to the utilities, whose only business was energy. Since the oil embargo, however, many industrial users fear that they might be faced with problems of fuel allocation or excessive cost.

6-2 PRIOR OSES EXPERIENCE

In the United States, approximately 1000 all-fuel generating systems have been installed over the years to provide electricity, heating, and air conditioning in large residential, industrial, and office complexes.[6] Interest in these systems has been rekindled as a result of the nation's growing energy problems.[7,8]

Regrettably, less than half of those earlier so-called "total energy" systems are functioning as they were intended.[9] For a variety of reasons, many have never lived up to expectations. Some systems were designed, installed, paid for, and then never used. Others are being used only for standby power instead of the primary power for which they were designed. Still others operate on-line but have never achieved the promised operating economics and efficiencies.

Still, the rising cost and growing unavailability of energy resources make it increasingly important to investigate the potentially high efficiencies that OSES' offer.[10,11]

And OSES' offer other benefits. Thermal pollution is reduced since, with the reuse of waste heat, there is less need to eject water from utility condensers at elevated temperatures into nearby streams, rivers, and oceans.

Another long-term benefit is that the probability of brownouts and blackouts is reduced through lower demands on limited utility-generating capacity.

A further boost has been recent governmental encouragement through various tax incentives which favor cogeneration. Finally, experts feel that many existing OSES installations can be modified to substantially improve performance.

The following major circumstances are also likely to affect the status quo:

1. Fuel costs will remain high.
2. Industrial equipment designed for oil and gas firing will be replaced or modified for coal firing where possible.
3. Federal and state tax rate, incentives, and regulations are likely to be used to promote energy policies that favor conservation.
4. Restructuring of utility rates to reflect time-of-day usage will increase.
5. There will be more equipment with improved efficiency capable of operating on a variety of energy sources.
6. Public awareness of energy use and a perception of the advantages of on-site generation will grow.
7. The example set by a number of European nations which have increased their use of OSES' will be felt. Approximately 29 percent of the electricity produced in West Germany comes from on-site industrial plants.

118 Energy Conservation in Buildings and Industrial Plants

8. Innovative approaches to on-site generation and storage involving renewable energy sources are expected to become available at competitive costs in the near future.

9. OSES' reduce utility investment risks, since new units can be constructed and installed rapidly to accommodate unanticipated demand changes.

10. OSES' designed for high-efficiency energy use or combined with utility power generation reduce air pollution levels.

11. The basic infrastructure is now in place for a major expansion of OSES equipment to accommodate projected growth needs.

12. Other valid economic, social, strategic and political reasons may arise for reversing the centralization of U.S. electricity production.

6-3 ANALYZING POTENTIAL OSES APPLICATIONS

The checklists included here cover the major steps of an on-site generation feasibility investigation for a large, energy-consuming facility. The discussion pertains to systems proposed for both (1) an existing facility which does not now have on-site generation and (2) a facility in the design stage.

The evaluation procedure involves three distinct phases: collecting data, analyzing data, and preparing a written report.[12]

The checklists in Figures 6-1 to 6-3 delineate the types of information developed in each of the three phases. Since individual cases vary considerably, the lists represent a general guideline and not a specific set of rules.

The development of a valid feasibility analysis is a complex undertaking. A meaningful analysis requires sound engineering and economic principles, orderly procedures and calculations, and, most importantly, the exercise of broad perspective and detached judgment.

Raw data, both current and projected, are collected through personal visits, correspondence, and telephone communications with the client, equipment manufacturers, utility suppliers, and other sources. Figure 6-1 illustrates specific examples of data collection activities.

Particular attention must be focused on the time coincidence of electrical and mechanical demands versus heat demands at various temperature levels. These demands should roughly parallel each other for a given time period.

Data acquired in the first phase are developed for two types of analyses: technical and economic. The technical portion consists of matching energy characteristics to select suitable equipment types and sizes for system requirements. Since waste-heat recovery is a primary criterion in an on-site power system, particular attention should be directed toward prime movers, waste-heat boilers, and associated recovery equipment.

Introductory meeting with client establishes project guidelines:
- Scope of study
- General background
- Confidentiality requirements
- Client's economic objectives
- Pertinent information available from the client's records
- Special requirements

Energy forms and characteristics:
- Electricity: Voltage, phase, frequency, power factor, DC power needs, etc. Special requirements, such as computer-room applications or large-motor starting.
- Work: Possible use of prime movers to drive mechanical equipment directly.
- Heat: Temperature-pressure levels and conveying medium (steam, hot water, air, etc.). Also application: For comfort, for process.
- Refrigeration: Temperature levels. Also application: For comfort cooling, product or process cooling.

Energy load information:
(These data are gleaned from past records or current measurements, if available, or deduced from data on existing or proposed energy-consuming equipment, drawings, specifications or other available sources.)
- Energy consumption
- Maximum rates of demand
- Load factors
- Profiles of energy demand versus time
- Seasonal and/or cyclic patterns of load demand

Energy availability and cost data:
- Possible fuels: Gaseous, liquid, solid, by-product, waste
- Energy sources: Utility-supplied electric energy, utility-supplied natural gas, commercial fuel suppliers, in-house fuel sources
- Energy-supplier interface: Possible energy use restrictions and curtailments, possible buy and/or sell arrangement with the electric utility, possible standby arrangement with the electric utility
- Past utility and fuel bills
- Utility and commercial supplier rate schedules and adjustment factors, taxes

Other cost data:
- Construction and equipment installation
- Operating labor
- Plant maintenance
- Miscellaneous consumables, such as lubricating oil and water
- Insurance
- Property taxes
- Income taxes
- Credits from sale of energy

Equipment data:
- Manufacturers' published literature
- Other publications
- Experience and records from previous installations
- Nameplate data from existing equipment

Fig. 6-1 Data collection from a variety of sources. (Milton Meckler, "Preparing for Onsite Power," *Power Magazine*, April 1975.)

TECHNICAL ANALYSIS

Prime-mover and heat-recovery equipment:
- Reciprocating engines with heat recovery from jacket-water and/or exhaust systems.
- Gas turbines with unfired or refired heat-recovery boilers. Consideration given to closed-cycle gas turbines for use with solid fuels.
- Steam boilers in conjunction with steam turbines of various configurations—noncondensing, condensing, auto-extraction.
- Combinations of different types of prime movers and heat-recovery equipment.
- Standby requirements: Equipment installation or energy purchased from the utility.

Auxiliary equipment:
- Supplementary boilers
- Cooling towers
- Pumps
- Cooling equipment: Mechanical—reciprocating, centrifugal, etc.; absorption; combinations of cooling equipment
- Use of prime movers to drive mechanical equipment
- Standby requirements

Buildings and other structures:
- Number, size, design, materials of construction, etc.

Primary energy input requirements:
- Fuel: Primary, standby, or supplemental, if required
- Purchased electricity

Environmental factors:
- Noise and vibration
- Air and water pollution

ECONOMIC ANALYSIS

Cost estimates:
- Initial investment: Construction, engineering, supervision, etc.
- Operating costs: Fuel, purchased electricity; operating labor, maintenance; miscellaneous consumables; property taxes, insurance; credits from sale of energy; other items
- Income taxes
- Interest on construction loans

Analysis procedures:
- Cash flow: Net present value, discounted rate of return
- Net annual owning and operating costs
- Other methods
- Computer economic analysis, if applicable
- Alternate financing arrangements

Fig. 6-2 Technical and economic analysis developed from collected data. (Milton Meckler, "Preparing for Onsite Power," *Power Magazine,* April 1975.)

6-4 SELECTING THE PRIME MOVER

Selecting the proper prime mover and heat recovery equipment is fundamental to the success of an on-site system. Factors that bear on proper selection are:

1. Size of the installation
2. Temperature levels of rejected heat needed to effect recovery
3. Relationship between heat and power requirements (heat-power ratio)

A fundamental criterion of feasibility is that the possibilities for utilizing rejected heat coincide reasonably, in time and quantity, with production of work. This is

Outline of subjects to be covered:

- Introduction, explaining purpose and scope of report
- Summary and recommendations
- Background information relating to reasons for undertaking the study
- Information sources and methods used to develop data included in the report
- Description of existing facilities, if any
- Current and projected energy requirements
- Possible energy sources
- Description of alternate systems considered, including discussions of cost, reliability, technical factors, fuel sources, pollution abatement and waste disposal
- Economic analyses of alternate systems considered
- Discussion of less tangible factors, such as future trends of energy availability and cost, plant-reliability experience, and degree of precision obtainable in projections
- Conclusions reached in the study
- Tables, diagrams and figures throughout the report, as required, to illustrate energy loads, costs, rate schedules, flow schematics, and preliminary physical layouts of proposed facilities
- Appendix

Fig. 6-3 Report stems from analyses. (Milton Meckler, "Preparing for Onsite Power," *Power Magazine*, April 1975.)

because of the difficulty of storing heat. Heat storage facilities are bulky and costly, and demands for work and heat rarely coincide exactly. Therefore, prime mover heat recovery systems almost always require auxiliary equipment to supply additional heat or to dissipate excess rejected heat.[3]

Figures 6-4 and 6-5 illustrate flow diagrams for energy systems using various types of prime movers and heat recovery equipment. The qualitative energy balance for each system is:

1. Energy input
   a. Fuel to prime mover
   b. Fuel to direct-fired boilers and other equipment
2. Energy output and losses
   a. Useful output
      (1) Power load supplied by prime mover
      (2) Heat load supplied by prime-mover-rejected heat or by direct-fired equipment
   b. Losses
      (1) Mechanical and electrical losses.
      (2) Exhaust gas or stack losses resulting from limitations on temperature levels of recovered heat or dewpoint temperature of gases. This includes hydrogen losses when the energy balance is based on fuel high heat value.
      (3) Radiation, transmission, and convection heat losses.
      (4) Unused prime-mover-rejected heat.

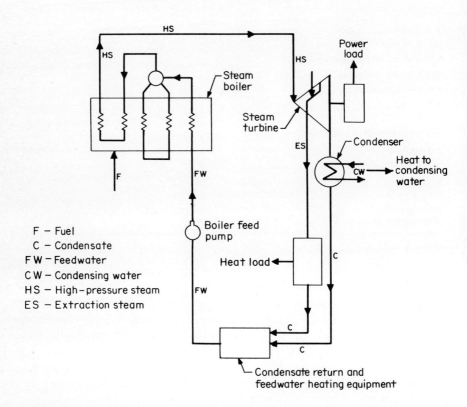

Fig. 6-4 Typical steam heat recovery system flow diagram for reciprocating engine. System can also be provided with pressure reducing and desuperheating if required. (Milton Meckler, "On-Site Energy Systems Stage a Comeback," *Electrical Consultant,* March 1975.)

Attention should also be focused on alternate methods of supplying energy needs (mechanical versus absorption cooling, for example, or electric motors versus prime mover direct drives), to optimize the demand balance between prime mover work and waste heat recovery.

Equipment selection greatly affects energy conservation. Systems can be all-electric, all-fuel, or combinations of the two. To satisfy load requirements correctly, many different combinations of equipment and systems must be considered. Mechanical refrigeration or absorption chilling may be used to satisfy the air conditioning load. A heat pump can be substituted for mechanical refrigeration. The output of a heat pump varies according to surrounding conditions; for a typical application,

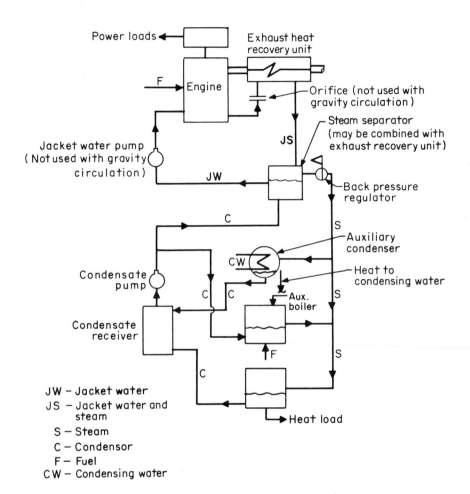

Fig. 6-5 Typical condensing: automatic-extraction steam turbine flow diagram. Note that auxiliary condenser can also use air as its condensing medium. (Milton Meckler, "On-Site Energy Systems Stage a Comeback," *Electrical Consultant*, March 1975.)

the coefficient of performance (COP) can vary from approximately 2 to 4. Also, due to a similarly high COP for mechanical refrigeration, its efficiency is much higher than any other component's efficiency.

After the selection of technically feasible system arrangements, equipment performance characteristics are compared to the system's energy load requirements. Thus, we arrive at primary energy usage over a cyclic period, usually a year.

The costs are developed into a basis for common comparison. It then becomes possible to see whether optimum costs are being achieved, and to determine the system arrangement that is most economically feasible.

6-5 PRIME MOVERS

The principle of heat recovery, fundamental to the operation of OSES', has been employed for some time. The passenger compartments of our cars are heated this way. Prior to World War II, the availability of systems for recovering engine heat was principally restricted to facilities large enough to justify the installation of steam turbines. Since that time, the range of feasible heat recovery systems has been extended to include users with much smaller power and thermal requirements. The typical OSES comprises a gas turbine generator or a diesel generator unit which utilizes waste heat to provide all or part of the space and domestic water heating, steam, and air conditioning needs.

The simple-cycle gas turbine is able to convert approximately 20 percent of the heat energy of the fuel to shaft horsepower. Typical operating gas turbine exhaust temperatures range from 600 to 1200°F (315 to 649°C). These are relatively high temperatures, useful for space heating, water heating, or the production of low-pressure steam which can be used to drive an absorption air conditioning system.[3]

A well-designed heat exchanger can recover over 85 percent of the exhaust heat from a turbine prime mover. Consequently, waste heat recovery systems operate with total thermal efficiencies in excess of 80 percent. The exhaust gases from the turbine are relatively clean and normally contain no measurable quantities of carbon monoxide, oil, or other pollutants. In past years, OSES' have been used in apartment complexes, shopping centers, educational institutions, and office buildings which require process steam or hot air for drying processes. Gas turbines for these systems can be designed for *dual fuels* (natural gas or diesel fuel oil interchangeably). The dual-fuels option allows a user to normally employ natural gas, often at more economical but interruptable rates, and still maintain system availability by switching to diesel fuel oil when necessary.

While brake thermal efficiency is low compared with other prime mover types, turbine drives are often more suitable for peaking service. Gas turbines are also used in combination with steam turbines in the combined-cycle arrangement discussed earlier. This allows for heat to be taken as steam from the heat recovery boilers or as extraction (exhaust) steam directly from the steam turbines. The brake thermal efficiency of a gas turbine can be improved through a regenerative cycle, in which the exhaust gases are used to preheat the combustion air.

The reciprocating engine, gas- or diesel-fueled, is often found in smaller projects, where space heating and cooling are the major application for rejected heat. Heavy-duty, low-speed diesel generator units are commercially available at power levels up to 6 MW. They are used primarily for on-site emergency power generation. A number of smaller, municipally owned utilities still employ such equipment for peaking service. Foreign manufacturers offer diesel-driven units rated up to 25 MW. Intermediate-speed, lighter-weight units are generally available at power levels up to 2 MW. Diesel engine units are generally arranged in multiples for base-load and for peaking-load operations. Although diesel generator heat rates range from 12,000 to 13,500 Btu/kwh (12,660 to 14,242 kJ/kwh), with heat rates comparable to a gas turbine generator at full-load conditions, diesel generators can maintain their design point efficiencies to reduced power levels. Diesels, however, require lighter, more expensive fuels than gas turbine units. Low capital cost, modularity, high response, and ease of maintainability are found with both types of units. However, the specific power in horsepower per pound (horsepower per kilogram) and power density in horsepower per cubic foot (horsepower per cubic meter) values for diesels are much lower. Consequently, most diesels are larger and heavier than comparable gas turbine units sized for equivalent power ratings.[13]

A computer analysis of energy-consuming and energy-producing equipment is recommended where practical. Feasibility depends on the quality of information available from equipment manufacturers.

6-6 THE ECONOMIC ANALYSIS

The economic analysis consists of estimating the initial investment and annual operating costs of each alternate system arrangement arrived at in the technical analysis. These costs are then developed into a common comparison to determine the most economically feasible system arrangement.

The overriding factor in such an analysis is the initial capital costs. For a specific energy system, the approach with the lowest overall cost may or may not also entail the lowest energy use. In most analyses, the lowest cost and lowest use of energy occur in the same system. Nevertheless, complex value judgments and trade-offs must sometimes be made.

Energy system costs can be divided into owning costs and operating costs. Owning costs are primarily a function of the construction or replacement price of the on-site facility. Some owning costs however, are not revealed as a straightforward cash outlay. These include such items as insurance, maintenance, taxes, and depreciation.

One complex consideration is the time value of money—the cost of borrowing money versus the immediate commitment of capital resources. A detailed review of the return-on-investment potential for capital that might be committed to an OSES should be undertaken. This value must then be compared with alternatives for borrowing capital and with overall savings expected through any projected energy generation system. The ownership costs of an on-site energy generation system are almost certain to be higher than those of any reasonable system that uses purchased energy. A more critical consideration is whether the consumption and cost of fuel is lower for on-site generation.[1,4]

When a determination is made that operating costs for an on-site total energy system are lower than the costs of a system using purchased energy, a complete analysis of various on-site approaches can begin. The first step in a complete analysis is to compile a list of annual cash flows for every year of the projected life of the particular system. Among the items that need to be included for each year are:

1. Investment costs for plant construction, both initial construction and additions or replacements in subsequent years.

2. Operating labor and supervision cost including fringe benefits and any applicable overhead.

3. Purchased fuel cost including primary, standby, and, if applicable, pilot fuels for prime movers, boilers, and other direct-fired equipment.

4. Purchased electric energy cost including demand, consumption, and other applicable charges.

5. Maintenance costs including parts and any labor not considered in operating labor.

6. Costs for miscellaneous consumable items such as water, lubricating oil, chemicals.

7. Cost of outside contracted services and supplies.

8. Cost of standby electrical service and anticipated charges for energy used standby periods.

9. Insurance costs.

10. Taxes.

11. Credit for receipts from net salvage value of items replaced or removed during economic life of project.

12. Credit for receipts from services supplied to energy consumers other than the primary energy-consuming facility.

13. Other costs or benefits resulting from the project.

Examples of annual cash flows are shown in the "actual costs" columns in Table 6-1.

The next step is to determine net present value of costs, net present value of savings, or uniform annual cost as illustrated by comparisons 1, 2, and 3, respectively, in Table 6-1.

The economic life of the project will ordinarily correspond to the project's physical life. Factors which may reduce economic life below physical life are technological obsolescence and the life of the serviced facility. The composite

TABLE 6-1 Cost Comparisons Based on Net Present Value

Comparison 1: Net Present Value of Costs

Project Year	10% Discount Factor	Actual Costs Option			Discounted Costs Option		
		A	B	C	A	B	C
1	0.954	250	600	700	238	572	668
2	0.867	250	100	75	217	87	65
3	0.788	250	100	75	197	79	59
4	0.717	250	100	75	179	72	54
5	0.652	250	100	75	163	65	49
TOTALS		1250	1000	1000	994	875	895

Comparison 2: Net Present Value of Savings

Project Year	10% Discount Factor	Actual Costs Option		Savings With B	Discounted Savings
		A	B		
1	0.954	250	600	350	334
2	0.867	250	100	150	130
3	0.788	250	100	150	118
4	0.717	250	100	150	108
5	0.652	250	100	150	98
TOTALS		1250	1000	250	120

Comparison 3: Uniform Annual Cost (Different Economic Lives)

Project Year	10% Discount Factor	Actual Costs Option		Discounted Costs Option	
		A	B	A	B
1	0.954	250	600	238	572
2	0.867	250	100	217	87
3	0.788	250	100	197	79
4	0.717		100		72
5	0.652		100		65
TOTALS		750	1000	652	875

Uniform annual cost:
Option A: 652/2.609 = 250
Option B: 875/3.978 = 220

SOURCE: Milton Meckler, "Preparing an Economic Analysis of an On-Site Energy System," *Consulting Eng.* (February 1976). Reprinted from *Consulting Eng.* Copyright © by Technical Publishing Co., a Division of Dun-Donnelley Publishing Corp. All rights reserved.

physical lives of plants generating heat and power can be estimated at 20 to 25 years for those with reciprocating engines or gas turbines, and 30 to 35 years for those with steam turbines.

The discount rate is another factor that varies. The minimum discount rate is the owner's cost of money (interest) over the economic life of the project. The degree to which the discount rate should exceed the cost of money depends upon many factors, including the risk or degree of uncertainty associated with the project.

6-7 PROJECTIONS INTO THE FUTURE

Evaluation of alternate energy supply systems involves projections into the distant future. Projected costs and savings near the end of the project life are inherently more error-prone than earlier projections.[4]

This effect is diminished because discounting causes early cash flows to have a greater impact than those occurring later in project life. Discounting decreases the effect of the cash flow estimates that are most subject to error.

Another uncertainty results from the fact that an energy supply system is only a part of a larger complex. Even though the supply system may function as predicted, unforeseen occurrences in the energy-consuming facility can alter system economics.

Information developed in the first and second phases must be incorporated into a written report. Ordinarily, a preliminary draft is prepared first, for the client's review and comment. The client's desired revision can then be incorporated into the final draft.

With the report in hand, the client should be able to evaluate energy options and arrive at a decision. Assuming on-site power is feasible, the client may show some reluctance to select this alternative because of high first cost. As energy costs soar, however, such decisions are becoming easier.[12]

Determining OSES feasibility requires a careful and thorough study of all available options. This demands substantial amounts of technical skill, time, and effort. However, with skyrocketing costs of fuel, OSES' are finding themselves paying out much faster than ever before.

6-8 THE NEED FOR COGENERATION

It has been estimated that approximately three-quarters of the energy used by industry actually performs useful work; the rest is waste heat. In addition, two-thirds of the energy used in electricity generation and distribution is wasted. In a recent year, waste heat from these sources equaled over 7 million barrels of oil per day.

One way to utilize this waste heat is through cogeneration systems, which simultaneously produce thermal and electrical or mechanical energy. Cogeneration provided 15 percent of U.S. energy as recently as 1950, but it now contributes only 4 percent. Although cogeneration is economical today and will become increasingly attractive as energy prices rise, a variety of institutional barriers have impeded its development. Now the National Energy Act (NEA) requires that utilities buy and sell power to and from qualified cogenerators at fair and reasonable rates. Special tax credits were discussed in Congress but were deleted in the final version of the bill.

Cogeneration systems are generally attractive in situations where there is a load demand for heat. Laundries, hospitals, and the food, paper, refining, and chemical industries are prime candidates.

Some applications also exist in factories which operate on a three-shift schedule to permit maximum utilization of the generating facilities. Many industries use electricity in ratios amenable to cogeneration and require steam at temperatures which can be conveniently supplied with cogeneration systems.

Refer to Fig. 6-6 for schematic diagrams of five basic cogeneration systems. They represent the options of current interest to both industrial users and electric utilities across the country. Basically, these systems are steam-topping cycles in which both process steam and electricity are produced.[2]

System *a* is representative of a user who owns and operates facilities for producing both electricity and process heat. This system does not interconnect with the utility grid and is, therefore, incapable of supplying any net energy to it.

System *a* may often require energy from the serving utility to make up the difference between what it can supply and its total needs at a given time. Although such systems tend to reduce the total utility loading, its beneficial or adverse effect on the utility depends upon the mix after the utility input, as well as the amount of user standby power needed. The user must supply the skilled personnel necessary to run the plant and be directly responsible for any pollution control equipment necessary.

System *b* is identical to system *a* except that the user is coupled directly to the electric utility grid and therefore capable of delivering it to any net energy available. Users with a high demand for process steam (or high-quantity heating) but a low electrical demand may find such systems economically attractive, because the amortization for OSES' can be significantly reduced by income received from the local utility. Unfortunately, the frequency of such net deliveries may be unpredictable, or the system may not be capable of supplying power when the utility is experiencing high load demands. In such cases, the amount the utility will pay for the electricity may vary. A further utility concern in interconnected operation is system stability.[14] Improperly controlled excitation systems can result in operating problems within the remaining system and can cause negative damping to system transient conditions. Overall utility system stability can actually be reduced during system disturbances. Such problems are usually overcome through close cooperation between the user and the serving utility during the design and construction phases.

The remaining systems, namely, *c*, *d*, and *e*, are cogeneration facilities owned and operated by the utility.

System *c* represents shared responsibilities: while the utility owns and controls

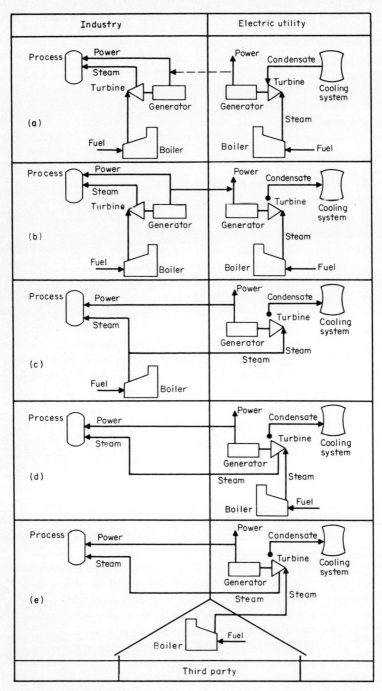

Fig. 6-6 Basic cogeneration systems. ("Cogeneration," *Power Engineering*, March 1978, p. 34.)

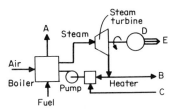

Fig. 6-7 Steam topping cycle: (A) stack, (B) process steam, (C) feedwater, (D) generator, (E) electric power. ("Cogeneration," *Power Engineering*, March 1978, p. 37.)

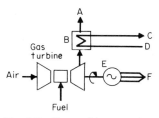

Fig. 6-8 Gas turbine topping cycle: (A) stack, (B) waste heat boiler, (C) process steam, (D) feedwater, (E) generator, (F) electric power. ("Cogeneration," *Power Engineering*, March 1978, p. 37.)

the generator, the user owns and controls the steam-producing equipment. Unfortunately, some states prohibit the sale of steam to all but utilities. And load-matching problems could result if the user reduces the steam supply because of plant operational or capacity changes.

System *d* represents the greatest investment for the utility. It does, however, afford it some flexibility, since it can be built around a large coal or nuclear steam-topping cycle (see Fig. 6-7), a smaller gas- or diesel-topping cycle (see Figs. 6-8 and 6-9), or a combined-cycle plant (see Fig. 6-10).

System *e* minimizes the user investment inherent in system *c*: a third party owns and operates the boiler. The problem here is to find three parties who would benefit from such an arrangement and be willing to make the necessary long-term commitments.

In addition to the steam-topping and combined cycles, cogeneration facilities can employ the steam- or organic-bottoming cycles illustrated in Figures 6-11 and 6-12.

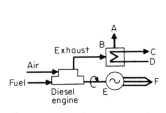

Fig. 6-9 Diesel topping cycle: (A) stack, (B) waste heat boiler, (C) process steam, (D) feedwater, (E) generator, (F) electric power. ("Cogeneration," *Power Engineering*, March 1978, p. 37.)

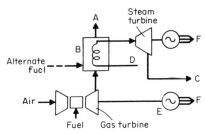

Fig. 6-10 Combined cycle: (A) stack, (B) waste heat boiler, (C) process steam, (D) feedwater, (E) generator, (F) electric power. ("Cogeneration," *Power Engineering*, March 1978, p. 37.)

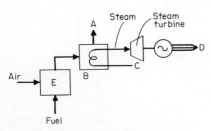

Fig. 6-11 Steam bottoming cycle: (A) stack, (B) waste heat boiler, (C) feedwater, (D) electric power, (E) furnace, incinerator, or fired boiler. ("Cogeneration," *Power Engineering*, March 1978, p. 37.)

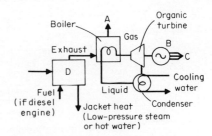

Fig. 6-12 Organic bottoming cycle: (A) stack, (B) generator, (C) electric power, (D) moderate- or low-temperature heat source. ("Cogeneration," *Power Engineering*, March 1978, p. 37.)

Table 6-2 provides a summary of the salient features, cost, and applicable sizes for each of the six cogeneration system types.

6-9 INDUSTRIAL COGENERATION POTENTIALS

Another productive use of waste heat from large-scale cogeneration facilities is now being fully explored: district heating. Because of favorable reports from Sweden and West Germany, state public utilities commissions will undoubtedly be paying closer attention to this option when they process applications for new generating capacity.[15]

A recent cogeneration study[16] of six major industries including food, textiles, pulp and paper, chemicals, and petroleum refining suggests that significant opportunities exist in the area of process heat and steam topping in applications of 5 MW and greater. Petroleum-refining and pulp and paper industries show the greatest potential growth, because existing cogeneration facilities are already serving the other industries. Under long-term contract, the chemical industry presently purchases the most steam of the six industries. The textile and food industries experience the greatest daily and seasonal steam demand variations, thereby reducing growth imports.

Other technical factors which tend to constrain wider industrial use of cogeneration include:

1. Need to supply steam drive mechanical.
2. Costs of voiding long-term steam purchase contracts.
3. Wide steam load fluctuations process.
4. Process steam demand is too low.
5. Waste heat boiler steam pressure is too low.
6. Utility standby charges for supplementary electric power are too high.

TABLE 6-2 Cogeneration Cycle Configurations

Cycle	Size, MWe	Fuel	Elect. Steam, kW/10⁶ Btu	FCP, Btu/kWh (kJ/kWh)	Process steam press, psig (kgf/cm²)	Total plant installed cost, $/kW	Pollution	Controls	General System Notes
Gas turbine and waste heat boiler	0.5 → 75	Gas No. 2 oil Treated residual SNG (low Btu)	200	5500 (5802)	150–600 (10.5–42.2)	350–400	NO_x	Water or steam injection	1000°F exhaust can be used as clean hot gas
Diesel engine and waste heat boiler	0.5 → 25	Gas No. 20 oil Treated residual	400	6500 (6857)	15–150 (1.1–10.5)	350–500	NO_x Particulates	Tuning Steam injection Baghouse	Efficient at part load and in small sizes High power-steam ratio
Steam boiler and turbine	>1	Nuclear Any oil Coal Wastes	45 → 75	5000 (5275)	15–600 (1.1–42.2)	500–600	SO_2 Particulates NO_x	Low S fuel, scrubber Precipitator	Efficient at part load
Combined cycle and waste heat boiler	1 → 150	Gas No. 2 oil SNG	150	5000 (5275)	15–900 (1.1–63.3)	350–450	NO_x	Water or steam injection	Variable power-steam ratio Back-pressure steam turbine
Steam bottoming	0.5 → 10	Waste heat	NA	0	NA	400–600	NA	NA	Efficient at part load
Organic bottoming	0.6 → 1	Waste heat	NA	0	NA	400–700	NA	NA	Efficient at part load Uses exhaust Prototypes available Requires cooling water

SOURCE: "Cogeneration," *Power Eng.*, p. 37 (March 1978).

In the final analysis, the growth of industrial cogeneration or on-site generation depends upon the perceived benefits to both the users and the electric utilities. Without doubt its development can be stimulated by government action involving some combination of tax credits, direct government loans, or loan guarantees and relaxed air pollution requirements.

6-10 FUEL AVAILABILITY VERSUS PURCHASED POWER COST

There are times when the decision to convert from purchased to OSES power must be delayed until a more favorable investment climate exists. It is often possible to establish a graphical model of OSES plant economics expressed as shown in Fig. 6-13. All subsequent changes to either the cost of purchased fuel or power can be compared to determine if a shift favors the status quo or a change. The example shown in Fig. 6-13 was taken from an actual case study. The user was considering the installation of an OSES for his facility, but only if it resulted in a return on investment exceeding 14 percent. A subsequent check made two years later following a major increase in utility rates suggested that conditions were at hand for such a change.

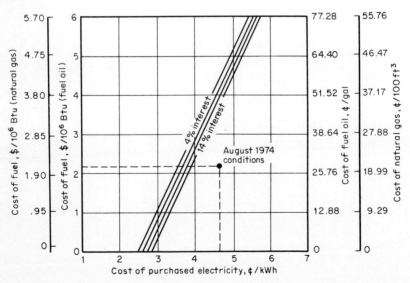

Fig. 6-13 Energy economics: illustrative power plant. (Milton Meckler, "Improving Performance of Existing On-Site Energy Systems," *Building Systems Designs*, February/March 1976.)

6-11 SOLAR SYSTEMS

To reduce our national demand for fuels, we must improve the efficiency of solar energy use. To do this we must concentrate on second law effects. A very large fraction of the fuel consumption of any building goes to provide low-temperature heat and uses, such as domestic hot water heating in the range of 110 to 120°F (43 to 49°C) and space heating in the range of 90 to 110°F (32 to 43°C). Consuming a high-quality fuel at 1800°F (982°C) or more to obtain low-temperature heat is fundamentally wasteful. Heating hot water by direct fuel combustion (assuming a 70 percent heat transfer efficiency) requires approximately eight times as much fuel as a reversible heat engine or a Rankine-cycle expander driving a heat pump would require to do the same job.

In short, work is intrinsically a more valuable form of energy than its equivalent Btu (kilojoule) content as heat. The value of fuel depends upon the extent to which its energy content can be converted to work. Once work can be obtained from it, the work can provide low-temperature heat. The availability principle can be used to identify inefficient practices in the transfer of solar energy to building HVAC systems. For example, in conventional oil- or gas-fired space heating boilers, the ratio of the thermodynamic availability required to that consumed would be less than 0.10. For electrically driven refrigeration equipment, the same ratio would be less than 0.12. By concentrating solar collectors, we are able to utilize solar energy at higher temperature levels to do useful work (such as to drive heat recovery chillers or heat pumps) and to store the remaining thermal energy sequentially at progressively lower temperature levels for building space and domestic hot water heating needs.

The utilization of solar energy by heat pump systems in large buildings falls into three general modes: (1) collecting solar energy at a temperature below that required by the heating application, and using it as a low-temperature evaporator heat source[20] (the so-called *solar-assisted approach*); (2) collecting solar energy at some temperature well above that required for the thermal load and using it to drive a heat engine–vapor compression heat pump (the so-called *solar-powered approach*); or (3) employing some combination of modes 1 and 2. A recent study has shown that solar-powered heat pumps are technically preferable to direct heat exchangers for the heating applications of absorption-cycle heat pumps.[17]

6-12 CONSTRUCTION FEATURES OF THE POLYPHASE INDUCTION MOTOR

The introduction of the polyphase motor represented a major breakthrough for highly mechanized industries. Where ruggedness and simplicity are necessary, the squirrel-cage motor is without doubt the principal choice for most constant-speed drive requirements. The wound-induction motor finds its principal application in situations where variable speeds, highly controlled torques, and/or low starting torques are desired. Its starter windings can be arranged in multiples of two to provide any reasonable number of poles. If we look at its full-load torque (FLT) versus speed as shown in Fig. 6-14 we find that, assuming a starting torque of 120 percent of FLT as it approaches synchronous speed, it rises to 245 percent at 85 percent speed. This is followed by a drop to FLT at 98 percent speed (representing a 2 percent slip). As we would expect, at synchronous speed, we develop no torque. However, as we drive the rotor, its synchronous speed, in the same direction as the motor, continues to draw a magnetizing component from the line but now delivers a power component. Thus it can be made to operate as an induction generator delivering a rated load, at a negative 2 percent slip and at normal lagging power factor.

The squirrel-cage rotor operates as a transformer with a shaft. Its locked-rotor torque, full-load slip, and locked-rotor kilovoltamperage principally depend upon

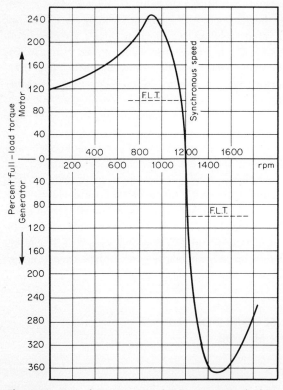

Fig. 6-14 Typical torque curve showing operation both as induction motor and as an induction generator. (Milton Meckler, "International Congress on Building Energy Management," Proceedings, Pergamon Press, London, 1980.)

the resistance and reactance of the (squirrel-cage) winding. Even if the rotor cage resistance is changed to maximum torque, it remains essentially the same, since, as in a transformer, a high-resistance secondary winding causes a relatively low locked-rotor kilovoltamperage.

The induction motor-generator has been with us since the first induction motor. Add to the induction motor a prime mover, an overspeed protection device, and possibly an accurate speed-sensing method, and the motor will be ready to generate. This one device has the ability to bring a load to speed and to drive it, to regulate the speed of another prime mover, or to provide a power system with real power. All these functions are available with a minimum of controls and a "couple it up and switch it on" type of installation. Where an external power system and a constant source of unused energy are available, the lowest first-cost, more reliable, and most maintenance-free energy recovery device providing electrical power is in the induction motor-generator.[18]

Upon connecting an induction generator to an alternating-current system, the system voltage and frequency are established at the machine terminals. The speed of the generator then determines the amount of generated current and its corresponding power factor. The generator has its design rated output, just as the induction motor has its nameplate loading. Operating at other than the rated load, rated voltage, or rated frequency carries with it all the problems of similar motor operation, such as high temperature, excessively high currents, and lower power factor.

The induction motor-generator operates in the overspeed range, above synchronous speed. Since the generator is coupled to a prime mover, it may be subject to further overspeeds. However, the generator manufacturer will normally design to the standard NEMA overspeed limits for induction motors. The induction generator can perform several duties in a string of equipment to which it is interconnected by means of a common shaft. For example, when generating, the induction machine can provide part of the building's power requirements. When the expander is unable to extract available energy or is down for repairs, the induction machine can drive the refrigeration compressor and thereby keep the power recovery string in operation. The induction motor-generator unit also provides the valuable function of speed controller.

Typically, the induction motor-generator will, for a limited time at least, produce up to twice its rated torque, or it will generate up to twice its rated kilowatts, thus providing or consuming torque and keeping its equipment train within $\pm 2\frac{1}{2}$ percent of its rated synchronous speed. The motor-generator does this automatically. No switchgear is required. The induction machine is inherently a load-matching device.[19]

Several firms now manufacture smaller expander induction generator sets for waste energy recovery.[20-22] The sets are rated roughly from 100 to 3500 hp and have been used mainly in air separation plants. As the gases pass through the expander, they give up heat energy, which is the power output of the expander.

The performance curves of a typical induction motor-generator are shown in Fig.

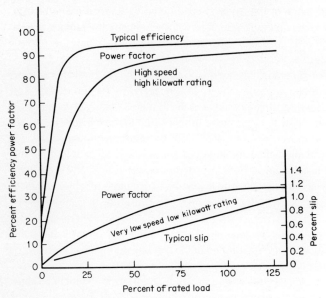

Fig. 6-15 Typical induction generator performance curves. (Milton Meckler, "Promising Solar Applications in Large Building HVAC Systems," *ASHRAE Journal*, November 1977.)

6-15. Although the efficiency normally remains high (about the same as a synchronous generator from rated load down to one-quarter load), the power factor drops off continually from near rated load. Thus, the range of economical operation is somewhat limited. If the unit has no starting requirements in motoring operation, the induction motor-generator manufacturer will normally install a low-loss cage winding, providing peak efficiency performance.

6-13 RANKINE-CYCLE POWER

Another promising technique now being evaluated utilizes a solar-powered Rankine cycle incorporating an induction motor-generator set for supplementing compressor needs on a demand cooling cycle. It can also produce electric power and store the excess by means of storage techniques such as flywheels and batteries.[23,24] Significant savings can result in the operation of a central chiller when its compressor is driven by a thermal engine arranged as a heat recovery machine or equipped with a double-bundle condenser. The Rankine-cycle fluid is indirectly heated by solar collectors.[19]

The Rankine-cycle engine derives its name from the closed cyclic sequence of thermodynamic processes experienced by a circulating working fluid. In its simplest form, the Rankine-cycle engine consists of a boiler, condenser, expander, and feed pump. Heat addition in the engine occurs at the boiler, which provides saturated or superheated vapor to the expander by transferring heat from the solar heated fluid to the Rankine-cycle working fluid. Power is extracted in the expander (either rotary, turbine, or reciprocating), after which the fluid passes on to the condenser, which supplies saturated liquid to the feed pump. The feed pump raises the pressure and resupplies fluid to the boiler, thereby completing the cycle. Heat is rejected from the working fluid in the condenser to an exterior cooling tower. Expander shaft work may be transmitted through a speed-reducing gearbox and used as a shaft power to drive refrigeration compressors or to drive an induction motor-generator to produce electrical power.

6-14 RANKINE-CYCLE ENGINES

Turbines have several advantages over reciprocating engines: simpler construction, smaller friction losses, no lubricant required in cycle fluid, low maintenance, high reliability, lower noise and vibration, easy regulations, and smaller size with large expansion to condenser pressure and temperature. Their major disadvantages are their cost and the fact that presently available small turbines have low efficiency (below 50 percent).

Studies[25] on small turbines driven by low-temperature solar energy have emphasized low turbine cost without too much concern for efficiency. More recently, a high-efficiency (77 percent isentropic) unit using organic fluids has been built and operated. Proper design and mass production techniques seem to be at the point where the cost of high-efficiency turbines can be reduced sufficiently to compete with reciprocating engines. Basic criteria in the selection of a working fluid for a solar-powered Rankine cycle[20] are the temperature level of the energy available from the solar collector and the condensing temperature. Chemical stability considerations generally limit maximum cycle temperatures for organic fluids to approximately the 400 to 800°F (204 to 426°C) range, depending on the fluid.

Organic fluids in Rankine cycles have some advantages over water. These include higher molecular weight and, possibly, positive slopes of temperature versus entrophy at the vapor mixture saturation line. The principal advantages of organic working fluid systems are:

1. Expander (turbine) efficiencies remain good despite the relatively low turbine inlet temperature of 600 to 700°F (315 to 371°C).
2. Low system operating pressures of 1.38×10^6 to 2.11×10^6 Pa, 200 to 300 psia (1,378,940 to 2,068,410 N/m²), may allow unattended operation.
3. Material compatibility of the system is good.
4. Low flame temperatures can be used in heating, resulting in low emissions of oxides of nitrogen (NO_x).

The development of the Rankine-cycle engine has advanced during the last few years. Applications of these developments have been evident in the low-power automobile and in military and remote base uses and with solar panels.

6-15 MODULAR INTEGRATED UTILITY SYSTEMS

To date, a number of significant studies, involving an assessment of energy saved and competitiveness with conventional systems, suggest that modular integrated utility systems (MIUS) can play a significant role in an energy conservation strategy for larger building and multibuilding community projects. MIUS are self-contained OSES' which can also provide major project utility requirements, including electricity, HVAC, water supply, solid waste, and wastewater management. They do this while consuming minimum quantities of fuel, water, or other natural resources. This is accomplished by novel means, including the reuse, or recycling, of materials and energy. In view of our need to supplement and eventually to replace nonrenewable fuel sources with renewable energy sources, recent interest in cogeneration[5] suggests a need to explore applications involving solar energy. Several unique advantages come to mind:

1. Potential for interaction between MIUS and solar thermal-power systems.
2. Potential for non-HVAC application, in solid waste, wastewater, and potable water subsystems, etc.
3. Mutually beneficial sharing of thermal storage and centralized solar collection.

When we consider the temperature levels associated with solar collectors and various HVAC and power subsystems, we find an interesting fit. Possible MIUS non-HVAC applications include wastewater influent temperature stabilization, sludge dredging, power generation for motors serving pumps, and other electrical needs. We shall not attempt to cover non-HVAC MIUS concepts in detail here, as there is much excellent work on the subject. Our point in introducing the general subject is to explore further the use of that class of heat engines known as Rankine-cycle engines. We hope to suggest some useful applications for converting solar heat to work by means of solar collectors that are available to operate within the temperature ranges for which experience does not exist.

In the following discussion we shall concentrate on three main areas of solar-heated, Rankine-cycle application:

1. Power generation.
2. Power generation incorporating conventional engine generator waste heat.

3. Power generation incorporating integration with building HVAC and other MIUS subsystems, including interface with power grid (i.e., cogeneration).

Figure 6-16 illustrates a typical, solar-heated, Rankine-cycle engine power generation system arranged to produce grid power consistent with equipment, collector, and land area constraints. An advantage of this approach is that, through interconnection with the MIUS, grid power can be shared by all interconnected motors or other auxiliary electrical equipment without the need for redundant motors or backup controls. For those periods of little or no sunshine, conventional diesel (after burn) engine prime movers are either activated or sized for the peak load condition. Automatic switching equipment can be provided to draw upon supplementary power from the reserving utility power grid to meet load demands.

Figure 6-17, though similar to Fig. 6-16, takes advantage of the heat from the conventional prime mover engine to also drive an adjacent organic Rankine-cycle engine through a bottoming cycle. By interconnecting the two prime movers in the manner shown schematically in Fig. 6-18, we achieve a hybrid waste-heat bottoming cycle. It is solar-augmented and can be combined to satisfy backup requirements by employing a conventional prime mover sized for roughly 75 percent of the peak load condition (assuming a prime mover efficiency of 38 percent). Automatic switching

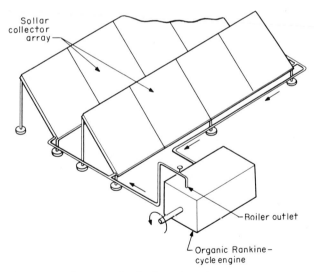

Fig. 6-16 Schematic of solar-heated engine power generator. ("Feasibility Study of Solar Energy Utilization in Modular Integrated Utility Systems," June 30, 1975. Prepared for the National Aeronautics and Space Administration, Lyndon B. Johnson Space Center.)

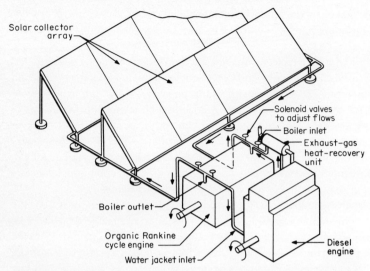

Fig. 6-17 Schematic of solar-bottoming Rankine power system. ("Feasibility Study of Solar Energy Utilization in Modular Integrated Utility Systems," June 30, 1975. Prepared for the National Aeronautics and Space Administration, Lyndon B. Johnson Space Center. Submitted by Arthur D. Little, Inc., Cambridge, Massachusetts 02140.)

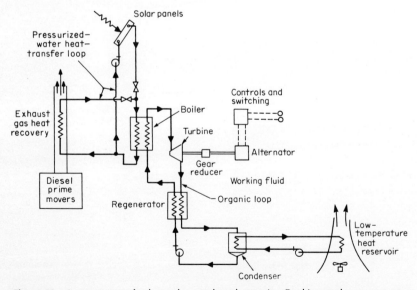

Fig. 6-18 Arrangement of solar and waste-heat bottoming Rankine-cycle power system. ("Feasibility Study of Solar Energy Utilization in Modular Integrated Utility Systems," June 30, 1975. Prepared for the National Aeronautics and Space Administration, Lyndon B. Johnson Space Center.)

equipment would also be needed to draw emergency power from the serving utility power grid in the event of the failure of either prime mover.

6-16 COMBINED-CYCLE PLANTS

For many years, the trend in electrical power generation has been toward greater cycle efficiency, principally through higher temperature, larger unit design, etc. Environmental laws and laws concerning site location (see Chap. 8) have resulted in a rethinking of this trend. Recognizing the potential of waste heat utilization, we now place a greater emphasis on integrated combined-cycle (refined) coal plants. Such plants offer a near-term energy solution and a possible intermediate solution which features improvement in the overall cycle while simultaneously reducing environmental impacts as indicated in Fig. 6-19.

If modular combined-cycle plants are to assume a greater share of base electric power generation, an improvement in cycle efficiency must be found to compensate for fuel conversion losses and lower thermal efficiency as compared with conventional, larger, high-pressure, steam-cycle generating units. One way of closing the gap is to employ clean fuels and thereby offset the penalty imposed by steady gas cleanup. Such fuels may take the form of hydrogen in the future, but immediate prospects point to coal-derived fuels and advanced high-temperature turbine systems approaching a 45 to 50 percent cycle efficiency.

6-17 HARNESSING ON-SITE GEOTHERMAL ENERGY

The rising cost of energy resources makes viable ECMs that have been possible for some time but have not been thoroughly developed because of abundant, low-cost energy. Sufficient economic incentive did not previously exist.[1]

Geothermal energy is currently receiving widespread interest in the United States as an available alternate energy source, particularly for electric power generation. *Geothermal energy*, as we define it here, refers to the thermal energy stored in deposits of dry steam, hot water, and hot dry rock. Although dry steam deposits are technologically the easiest to exploit, they occur only about one-twentieth as often as common hot water deposits. While there may be immense amounts of energy stored in hot dry rock, the technology for recovery of this energy has not been proven. Nor has significant exploration been carried out to determine if and where exploitable deposits exist. Consequently, during the next decade, geothermal energy development in the United States will likely emphasize exploitation of the water-dominated resources.

While these resources represent a significant energy supply, their rate of development will be affected by two basic factors: (1) chemical conditions encountered which will likely require special conversion systems and (2) economic considerations which require the development of conversion machines with the highest efficiencies to minimize the number of wells per unit of electrical capacity. An accelerated effort is currently underway nationally to support a broad-based research and development program.

One attractive energy source is the subterranean geothermal energy potential of various areas of the western United States. While some of these have been developed, the resource remains largely untapped.

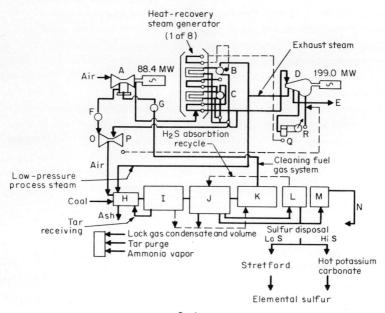

	System	
	Lo S	Hi S
Gas turbines (MW)	707.0	678.4
Steam turbines	199.0	198.3
Auxiliaries	21.0	21.3
Net plant Output	885.1	855.4
Heat rate (Btu/kW rated per hour of heat value)	10,300	11,100

Fig. 6-19 Schematic of advanced combined-cycle power plant: (A) MS-7001 gas turbine (one of eight); (B) low-pressure drum; (C) high-pressure drum; (D) steam turbine; (E) low-pressure process steam (reboilers, etc.); (F) intercooler; (G) gas heater; (H) gasifiers; (I) tar and particulate separation; (J) fuel gas cooling, ammonia and hydrogen sulfide absorption; (K) fuel gas saturator; (L) hydrogen sulfide absorber-regenerator; (M) ammonia strip; (N) ammonia to incinerator; (O) boost compressor; (P) steam turbine; (Q) below-flash-point deaerator; (R) intercooler. (John Papamarcos, "Combined Cycles and Refined Coal," *Power Engineering*, December 1976.)

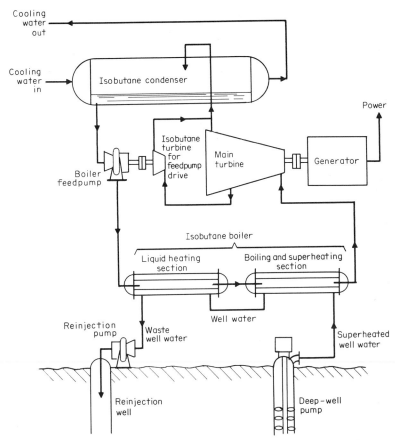

Fig. 6-20 Binary cycle system. (*Energy Fact Book, 1977*, Tetra Tech, Inc., Arlington, Virginia, April 1977, AD/A-038 802, p. XIV-12.)

To date, approximately 1000 MW of electric generating capacity utilizes geothermal powers. These facilities generally utilize dry steam, for which the technology has been proven. Since presently known sources of dry steam, such as the Geysers in northern California, are limited, the potential for providing significant supplies of energy is believed to rest with hydrothermal systems. Such formations often require innovative conversion and brine extraction[26] systems. Such systems broaden the economic base for using joint flushed-steam and associated mineral recovery and desalination values. This is crucial because of the problems normally associated with the more hostile geothermal brines, such as those found in the Imperial Valley of California. The use of a flashed-steam process is shown in Fig. 6-20. The binary cycle is illustrated in Fig. 6-20, and total flow power extraction concepts are shown in Fig. 6-21. Exchangers[27] are now in a developmental stage, but special problems must be resolved before commercialization can occur. Refer to Fig. 6-22 for the comparative net power extraction values of the three approaches described above, as a function of reservoir temperature.

One reason for the dearth of development in this area is the lack of definitive guidelines. Determining the feasibility of even preliminary exploratory efforts in a particular location can be problematic. Well drilling is, of course, necessary to determine the actual geothermal potential. However, even after reserves are proven, questions of energy availability, costs, potential markets, and other factors remain to be answered.

Developers hesitate to undertake the cost of an exploratory drilling program without reasonable assurance that any resources discovered will have a value sufficient to warrant the risk.

Regardless of the technical approach utilized, we must reach agreement on practical methods of determining the economic feasibility of commercial development once test drilling reveals the presence of adequate geothermal deposits.

Another possible alternative investigated, in one instance, was the sale of electric energy to adjacent utilities. Preliminary discussions with local utilities indicated they were willing to pay 4 mills/kWh for dump power and 7 mills for firm power. Since these values were substantially lower than the estimated cost (5.5 mills) and the total cost (11.5 mills) of geothermal power, this approach was abandoned.[26] However, more precise determination of the costs of geothermal power and further negotiations with utilities should be explored, since this may lead to more equitable arrangements in the future.

Power is, of course, only one of many possible commodities available from geothermal resources. The shortage of fresh water in an area can lead to a substantial demand for desalination facilities, which also require heat to operate.

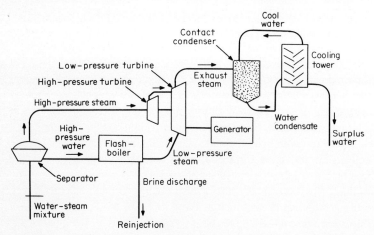

Fig. 6-21 Geothermal power plant. (*Energy Fact Book, 1977*, Tetra Tech, Inc., Arlington, Virginia, April 1977, AD/A-038 802, p. XIV-11.)

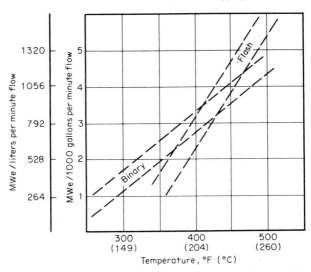

Fig. 6-22 Energy from given well flow. (*Energy Fact Book, 1977*, Tetra Tech, Inc., Arlington, Virginia, April 1977, AD/A-038 802, p. XIV-8.)

Geothermal development in the Imperial Valley, for instance, would require large amounts of water, but it could supply water as well.[27] This possibility should be considered in any discussion of the nonelectric potential of geothermal energy.

Another possibility to be explored is the recovery of valuable minerals from the geothermal deposits. This must be weighed against the associated material handling and disposal problems, and the need for reinjection.

The major cost of water desalination using geothermal energy lies in the fixed charges on heat exchange equipment. Since the size of the equipment is largely dependent upon the temperature of the steam, estimates of cash flows arising from desalination must await the outcome of exploratory drilling.

Geothermal recovery processes hold the greatest promise where saline or brackish water is available. The water desalination of geothermal fluids, whether occurring naturally or resulting from hydrocracking, is the key to major geothermal development.

6-18 RELIABILITY OF OSES POWER GENERATION

One way to improve the reliability of OSES' is to purchase a sufficient number of redundant units. Achieving high reliability can prove costly. Therefore, some interconnection with the utility power system to provide emergency power in the event of system failure is usually recommended. Providing utility backup for total electrical service requirements delivered by OSES' (advocated in the late fifties and sixties) may be impractical. Since the utility cannot release to the reserve generating capacity for other uses, the utility usually requests a high connect charge. Another possibility is the development of simple, mass produced, heat engines such as the Rankine, Stirling, free piston, and Ericsson. These operate with high reliability, compared to freezer and refrigerant air conditioners.

There is a paucity of statistical information on the performance of on-site generating systems. In part, this is because most installed systems serve only as emergency power systems. Furthermore, these systems tend to vary widely in design, control, site conditions, and level of maintenance. A summary of available data is given in Table 6-3.

In general, similar piston engine generator sets experience a mean time between failures of approximately 500 hours,[28] whereas diesel engines and gas turbines in the 50 kW to 1MW size range average 1.2 years between failures. Gas turbines have been known to run unattended satisfactorily for as long as 2.3 years between overhauls. The reliability of diesel-driven systems and the cost of system overhauls can be significantly improved by reducing engine speed to the vicinity of 1200 to 900 r/min for continuous operation. Most engines are normally operated at 1800 r/min for rated emergency power duty.

The widespread use of OSES' could reduce the overall cost of centralized utility power, since a lower system reliability could be tolerated. This is provided customers would accept lower reliability during peak demand hours and maintenance cycles.

TABLE 6-3 Reliability of On-site Equipment

Engine Type	Forced Outage Rate, %	Scheduled outage rate, %	Overall Availability	Mean Time between Failures	Average Repair Time, h	Reference
Diesel and gas turbines	1	3	0.96			1
Piston engines				500 h	2.5	2
750-kW gas turbine				700 h	6.4	2
5-MW gas turbine				1.8 yr	23.5	2
Large marine diesels			(More than 0.96)			3
Large diesels	1–4*			1.2–1.3 yr		4

*Assuming an average repair time of 100 h.
SOURCE: "MIUS Technology Evaluation: Prime Movers," U.S. Department of Commerce Report No. ORNL-HUD-MIUS-II, April 1974.

Under such circumstances, a utility grid backup for total service might also be acceptable to operators of on-site energy plants to cover occasional outages or down time for routine maintenance.

6-19 MAINTENANCE OF ON-SITE ENERGY SYSTEMS

Many industry managers considering the construction of OSES' remain reluctant to own and operate such systems, which are often regarded as complex. Uncertainties arise in projecting maintenance costs, personnel requirements, and system reliability. Operators for most OSES' require special training and licensing, and plant operation is subject to codes and regulations.

Actual maintenance and operational requirements depend on several variables. The minimum number of personnel needed, for example, will be dictated by design and operating conditions.[29]

Design conditions, such as distance or physical obstructions between components in the system, might require separate monitoring or separate attendance by operators.

The degree of automation of the facility will also affect the size of the work force. An underautomated plant will require a great deal of operating labor, while an overautomated plant may risk control malfunctions and minor control problems. While these can also result in excessive operating costs, there is usually a trade-off between initial dollar investment and long term operating costs.

Operating conditions, such as local boiler codes and the hours of operation, must also be evaluated.

In many locales, the safety code for steam boiler operation requires full-time attendance and certain boiler temperature and pressure conditions. Such regulations vary greatly from one section of the country to another.

If the system serves an installation where continuous, highly reliable on-hour operations are essential, the system must remain under continuous attendant operation.

Most systems require at least one or two operators. A plant designed for one operator features controls and monitoring instruments clustered for easy accessibility. The disadvantages here are personnel safety and the inconvenience of sickness, tardiness, and similar problems. As a result, such operations are generally limited to very small plants serving relatively noncritical facilities.

Two operators per shift allows one to perform adjustment tasks while the second operator serves as a backup. As time allows, the second operator may also perform some maintenance work.

In a plant operating continually, a single-operator system will require a minimum payroll total of six; a double-operator system will require a minimum of ten. This includes allowance for vacations, holidays, and sick leave.

A plant running 16 hours a day requires a payroll of four for a single-operator system; a double-operator setup requires seven.

An unattended plant or completely automated system is, of course, very rare. It would require a payroll of one for inspection and other minor chores.

Regular maintenance personnel are not included in these minimum payroll requirements.

Each plant, of course, would have a chief operating engineer plus at least one master or journeyman electrician and one journeyman mechanic. The need for other skilled personnel depends on the size and complexity of each system. Normally, the electrician and mechanic can handle all work within an energy plant. A plumber, pipefitter, or a sheet metal worker may be required for specific modification or repair work.

Continuous monitoring of all systems is necessary, regardless of the degree of automation. Data should include the local temperatures and pressures of engines and other equipment, generator electrical outputs, and the conditions of supporting equipment such as cooling systems, pumps and fans.

On a daily basis, data on lubricating oil levels, and fuel and chemical use, must also be collected. Day-to-day adjustment of such things as chemical concentrations, equipment efficiencies, and lubricant levels can be done at this time.

Weekly, biweekly, and monthly records must be kept. A one-week record on engines will provide an indication of the need for cleaning filters and other engine support components. Such records should also indicate any overlapping of operating schedules.

In addition to these regular data collection procedures, a system of reporting which allows management to perform efficiency studies and maintain a continuous check on plant operation must be developed. A log which serves this purpose is illustrated in Table 6-4. Great care must be taken in designing these log and data sheet forms to minimize time and effort of the operators. These sheets must also be designed to allow a simple and meaningful interpretation of the material. Ultimately, of course, all completed data and log sheets should be systematically filed to maintain continuous records.

These records will help prepare an effective preventative maintenance program, which is especially critical for dynamic equipment like that in a total energy system. Recorded maintenance costs from established plants and systems vary tremendously; this can be attributed directly to the presence or lack of proper preventive maintenance.

A preventative maintenance program can be easily set up. Each item system or piece of equipment must be completely detailed to determine its requirements. A complete manual should be written, tailored precisely to the individual plant and its equipment. It should describe all scheduled maintenance work required, as illustrated in Fig. 6-23.

Like other factors, maintenance costs vary dramatically from one installation to another. Costs for turbines, for example, include general service inspections, repair

TABLE 6-4 Service Log (Waukesha Engine)

Date: _____ Time: _____ Inspector: _____
Unit Hours: _____ Load (%): _____

		M	T	W	T	F	S	S
JACKET WATER	H.X. water press IN							
	H.X. water press OUT							
	H.X. water temp. IN							
	H.X. water temp. OUT							
	Jacket water temp.							
	Jacket water level							
LUBE OIL	H.X. water temp. IN							
	H.X. water temp. OUT							
	Oil press., engine							
	Oil press., filter IN							
	Oil press., filter OUT							
	Oil temp.							
	REN reading—oil consumed							
Gas pressure LEFT								
Gas pressure RIGHT								
Vacuum LEFT								
Vacuum RIGHT								
Engine exhaust temperature								
HRM exhaust temperature								
HRM steam pressure								
HRM steam temperature								

NOTES:
SOURCE: Milton Meckler, "Operation and Maintenance of On-Site Energy Systems," *Main. Eng.* (November 1975).

152 Energy Conservation in Buildings and Industrial Plants

Lubrication:
- Check engine oil and water levels daily and add oil if necessary.
- Check oil temperature daily. It should be between 180 and 220°F. If it's abnormally high or low, notify engine mechanic. Water temperature should be between 160 and 180°F.
- Measure oil pressure daily and make sure that it's between 40 and 50 psi at engine-governed speed and normal operating temperature.
- Governor—check oil daily and change after every 500 hours of operation. It must be changed every six months regardless of hours of operation.
- Air starter—inspect lubricator daily and keep filled to level with SAE 10 motor oil. Don't overgrease.

Drive belts:
- Inspect monthly for wear and adjustment.

Oil cooler pump:
- Service—Start chiller once a week to keep pump free. If pump is frozen, the drive belt will begin to smoke when engine is started. If this occurs, stop engine immediately. Take end off the pump and clean impeller and face plate. Make sure pump turns freely.

Air cleaners:
- Inspect air cleaner condition indicator gauge daily for airflow restriction. It's necessary to check only engines that are running.

Engine failsafes:
- All must be checked for proper settings and operation by engine mechanic every three months.

Gearbox to chiller coupling:
- Inspect coupling semiannually for alignment. At the same time, check bolts for tightness.

Fig. 6-23 PM plan for typical reciprocating engine. (Milton Meckler, "Operation and Maintenance of On-Site Energy Systems," *Maintenance Engineering*, November 1975.)

of minor damage and cleaning air compressors. Frequency depends greatly on the history of each unit and operational characteristics such as base load and peaking.

Maintenance costs for reciprocating engines, include checking lube oil levels, changing lube oil, engine tune-ups, and checking operating characteristics and performance. There are also costs for cleaning, performing minor repairs, packing bearings, replacing gaskets, and checking for unusual wear.

An analysis of these costs must take into account elements such as engine size, the running capacity factor, lube oil consumption, the type of fuel used, and the size of the unit.

Some of these costs can be controlled by using monitors, tachometers, various pressure and temperature gauges, and injection-pressure testers.

6-20 SITING ADVANTAGES

Numerous utilities are experiencing problems in locating suitable sites for power-generating facilities because of proliferating new regulations and environmental constraints which have materially increased both the time and expense required to justify adoption of a proposed site. (This phenomenon will be covered in more detail in Chap. 8.) Filing a comprehensive environmental impact statement (EIS) presents an opportunity for various environmental and other interested organizations to oppose new plants. OSES' require some of the same procedural steps as larger plants, but installation of several smaller plants contiguous with local industry and area development may be more acceptable than a remote site with a high local concentration of pollutants and an extensive power transmission network. Factors that favor OSES' over large utility generation plants include:

1. A relatively smaller impact on local environments.
2. In OSES', particularly those arranged for cogeneration, less energy is required for combined heating, industrial, and electrical needs. This results in a smaller incremental impact traceable to producing electric power generation.
3. Construction of a multiplicity of OSES' does not require significant dislocations to population or terrain. It avoids constructions of temporary labor camps, additional roads, or waterways.

Most alternatives to OSES' result in more adverse environmental impacts. Furthermore, some of the more grandiose schemes for large solar electric power systems would require extensive sections of land in a centralized area for the location of collectors or large towers. They would also need extensive transmission facilities. Smaller OSES' of the type described in this chapter could be integrated into buildings, adjacent parking structures, or manufacturing plants in the vicinity of the area to be served. This affords a lower public profile and, therefore, potentially less public objection and interference.

DESIGN CONSIDERATIONS FOR ON-SITE ENERGY SYSTEMS

In analyzing the potential for an OSES, the designer must collect data, analyze the data, and prepare a written report. Factors which bear on the proper selection of a prime mover include size of the installation, temperature levels of rejected heat needed to effect recovery, and the heat-power ratio.

A careful economic analysis complete with projections into the future should accompany any system proposal. Cogeneration systems, solar systems, and geothermal energy sources should be studied where feasible.

SUMMARY

Increasingly, on-site energy systems (OSES') are becoming important to industrial, commercial, and institutional users. Some systems provide total energy for the facility in question, while other systems utilize waste heat or other available energy sources to reduce the electrical load.

In deciding whether an OSES is feasible, many factors should be taken into account. A meaningful analysis requires sound engineering and economic principles, orderly procedures and calculations, and sound judgment. The selection of the prime mover and a thorough analysis of the economic viability of a given system must be weighed carefully.

Cogeneration will increase in importance in the years to come. Significant opportunities exist in industry for better utilization of process heat and steam topping.

A greater application of solar power in on-site systems, particularly in heat pump systems, will be needed in the future as well. Other innovations likely to improve the prospects for cogeneration include polyphase induction motors, solar-powered Rankine-cycle systems, and combined-cycle coal plants.

Where feasible, geothermal energy sources should be investigated. Sizable risks exist in the drilling stages and in forecasting the economic viability of such systems, but substantial benefits may accrue.

Special care must be exercised in the maintenance of on-site generation systems. Sound procedures should be established so that on-site facilities continue to run smoothly.

NOTES

1. M. Meckler, "Improving Performance of Existing On-Site Energy Systems," *Build. Sys. Des.* (February/March 1976).
2. M. Meckler, "Are Utilities Taking a New View at On-Site Power Generation?" *Electr. Consultant* (August/September 1976).
3. M. Meckler, "Options for On-Site Power," *Power Mag.* (March 1976).
4. M. Meckler, "Preparing an Economic Analysis of an On-Site Energy System," *Consulting Eng.* (February 1976).
5. "Cogeneration," *Power Eng.*, pp. 34–42 (March 1978).
6. "Single Energy Source Systems," SPAC Conference Proceedings, Oct. 10-12, 1966, Reinhold, New York.
7. "The Second Coming of On-Site Power Plants," *Buildings* (November 1974).
8. "The Little Engine That Seared Con Ed," *Fortune*, Dec. 31, 1978, p. 80.
9. M. Meckler, Guest Feature, "A Marketing 'Blueprint' for the New Energy Era," *IEEE Trans. Ind. Appl.* **1A-1Z**(3) (May/June 1976).
10. M. Meckler, "How Energy Costs in Commercial Buildings Can Be Cut 20 to 40%," *Electr. Consultant* (July 1975).
11. Gordon W. Neal, "Total Energy for Industrial Plants," *Consulting Eng.* (October 1976).
12. M. Meckler, "Preparing for On-Site Power," *Power Mag.* (April 1975).
13. "Report on Total Energy Feasibility Criteria," sec. 3, 1973, by OES, Inc., Long Beach, Calif.
14. *Stability of Large Electric Power Systems*, IEEE, New York, 1974.

15. R. Hohl, "District Heating in Switzerland," *ASHRAE J.* (March 1974).

16. "A Study of Inplant Electric Power Generation in the Chemical, Petroleum Refining, and Paper and Pulp Industries," June 1976, U.S. Department of Commerce No. PB-255-658.

17. H. M. Curran and M. Miller, "Comparative Evaluation of Solar Heating Alternatives," Hittman Associates, paper presented at Second Southeastern Conference on Application of Solar Energy, Baton Rouge, La., Apr. 21–23, 1976.

18. U.S. Patent No. 4,024,908.

19. M. Meckler, "Promising Solar Energy Applications in Large Building HVAC Systems," *ASHRAE J.* (November 1977).

20. Judson S. Swearinger, "Engineer's Guide to Turboexpanders," *Hydrocarbon Process.* (April 1970).

21. Judson S. Swearinger, "Turboexpanders and Expansion Processes for Industrial Gases," Cryotech 1973 Proceedings.

22. Judson S. Swearinger, "Turboexpanders and Processes That Use Them," *Chem. Eng. Prog.* (July 1972).

23. S. S. Penner and L. Icerman, *Energy, Volume II, Non-Nuclear Technologies*, Addison-Wesley, Reading, Mass., 1975.

24. L. A. Simpson, I. E. Adakar, and J. Stermscheg, "Kinetic Energy Storage of Off-Peak Electricity," Report No. AECL-5116, Whiteshell Nuclear Research Establishment, Canada, September 1976.

25. M. Meckler, "Cost Effective Solar Augmented Heat Pump/Power Building Systems," *Inst. Environ. Proc.* (April 1977).

26. M. Meckler, "Practical Guidelines for Geothermal Energy Development," *Geothermal Energy*, October 1975.

27. "Energy Factbook, 1977," U.S. Department of Commerce Report No. AD/A-038 802, April 1977.

28. "MIUS Technology Evaluation: Prime Movers," U.S. Department of Commerce Report No. ORNL-HUD-MIUS-II, April 1974.

29. M. Meckler, "Operation and Maintenance of On-Site Energy Systems," *Main. Eng.* (November 1975).

Industrial Sector Conservation Opportunities

Increasing the efficiency of energy and material use in our society is a relatively quick, relatively inexpensive and often an environmentally favorable alternative to the extraction, processing and use of more energy and materials.

John Quarles, Deputy Administrator,
Environmental Protection Agency

"Too much and yet too little" often describes the energy data currently being collected and published. While much information is published in the industrial sector, little of it relates directly to the needs of analysts, planners, and policy makers. Most analysts must deal with the casual relationship of changes in the energy system. Direct reporting of, for example, the total tons of steel produced in Pennsylvania in a given year (data normally collected by federal statisticians) will be of little help. We lack a set of standard definitive concepts and conventions to make compatible whatever data are collected. A national accounting system could provide a consistent framework for data collection.

Over the past 200 years, U.S. manufacturing firms shifted from wood to coal, then to oil and gas. Oil and gas currently accounts for roughly 78 percent of our primary energy consumption. Until recent years, the availability of low-cost energy resources placed few constraints upon our primary energy consumption and economic growth. Now domestic energy supplies are not keeping pace with our expanding requirements or with historical growth trends. If we are to become more energy-sufficient, coal will certainly have to play a substantial role.

7-1 INTRAINDUSTRY FUEL SUBSTITUTION POTENTIALS

Data gathered during recent studies[5] suggest that significant potential exists in the industrial sector for shifting from oil and gas back to coal. There are, however, significant near-term and long-term problems. Near-term difficulties include:

1. Lack of continuity in energy policies.
2. Lack of incentive policies for coal substitution.
3. Lack of proven technology for sulfur removal.
4. Lack of available low-sulfur coal.
5. Restrictive environmental regulations (opposition to strip mining, federal and state improved air pollution controls, etc.).
6. Supply and logistics problems, such as severe limitations on transportation facilities used to move the required tonnages over long distances to market.
7. Limited water supply.

Some of the long-term problems are:

1. Lack of a consistent national energy policy.
2. Lack of stabilization in the nuclear energy industry.
3. Lack of proven technology in coal gasification and liquefaction processes.

Some of these problems may be addressed by parts of the new National Energy Act, but many difficulties are sure to remain.

Although conversion is generally feasible for most manufacturing processes, certain industries require gaseous fuels under present and foreseeable technology for soaking pits, reheat furnaces, etc. While these specific uses do not significantly affect the total national fuel consumption, they are sensitive to short-term interruptions.

There is a method for estimating the price at which coal, oil, and gas would have to meet for there to be significant interest in conversion to coal for a given industry. As an example, we will use the iron and steel industry. The method consists of two principal steps:

1. Construction of some representative curves which portray energy substitution by fuel versus prices for the probable range of interest.
2. Construction of a curve of energy substitutions by fuel versus time.

In Fig. 7-1, we have illustrated two curves, based on two separate coal prices, nominally $A and $B/ton coal. Figure 7-2 provides a graphical method for comparison of the Btu (joules) to be shifted from oil to coal (substituted) for ratios of A/S, where X and Y represent two different fuel costs, expressed in dollars per barrel (dollars per cubic meter). Employing this approach, additional time-line data for oil and gas are accumulated, and, together with the data already represented in Fig. 7-1, the curves illustrated in Fig. 7-2 can be constructed.

Such curves are approximate; they represent only an appraisal of economic and technological factors which will determine the full time frame and extent of fuel substitution under given pricing assumptions.

Industrial Sector Conservation Opportunities   159

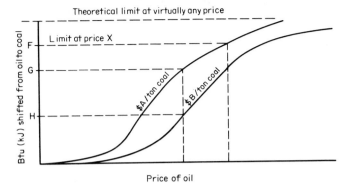

Fig. 7-1 Analysis of energy substitution by fuel versus price. ("Intra Industry Capability to Substitute Fuels," Federal Energy Administration, report no. FEA-E1-50034, October 1974.)

Once a family of such curves is developed, the amount of substitution at different prices for coal and oil can be determined for specific years of interest (for example, L and M in Fig. 7-2). A similar procedure is followed for evaluating gas substitution potentials.

Beginning at the left-hand side of the Fig. 7-1 curve, notice that, at relatively low oil prices, little if any substitution takes place. However, at a constant coal price, some substitution takes place as the price of oil increases. As the price of oil continues to rise, the curve slope also increases, suggesting a higher rate for substitution, until a point of saturation is reached at an asymptomatical limit corresponding to an infinite price of oil. If we assume that such relationships are inherently

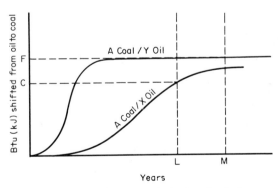

Fig. 7-2 Analysis of energy substitutions by fuel versus time. ("Intra Industry Capability to Substitute Fuels," Federal Energy Administration, report no. FEA-E1-50034, October 1974.)

160 Energy Conservation in Buildings and Industrial Plants

S-shaped, then we have only to fix the end points and compute some intermediate points for each curve.

Each industry has theoretical limits of substitution. An understanding of the economics of the various technologies is essential to fix the oil or gas prices associated with specific coal prices that would make substitution with that technology cost-effective. Once these data are gathered and Fig. 7-1 relationships are developed, we must estimate the time required for such technologies to be introduced. This is particularly important where additional development is required for commercialization.[1]

Referring to Fig. 7-1 and assuming an infinite amount of time for substitutions to take place, notice that, by selecting a point on any of the two curves, we automatically fix the amount of substitution taking place at given oil and fixed coal prices. By plotting that maximum substitution potential on a time line, it is possible to determine the extent of substitution that would take place at those fixed prices over a specific number of years.

7-2 THE VALUE-ADDED PRINCIPLE

Of increasing importance to the manufacturing and chemical industries is the amount of energy consumed per dollar of value added. In 1967, for example, the energy value ratio per 1967 dollar of value added for manufacturing and chemical sector, taken as a whole, approached 56,000 Btu/$ (59,080 kJ/$). Since the average U.S. energy cost to this sector in 1967 was approximately 65 cents/10^6 Btu (62 cents/10^6 kJ), this sector must have consumed roughly 3.6 cents of energy for every dollar of value added. By 1971, however, the ratio of energy consumed per dollar of value added ($VA) had fallen to approximately 53,000 Btu/$VA (55,915 kJ/$VA). Since the 1971 average energy cost rose to roughly 79 cents/10^6 Btu (75 cents/10^6 kJ), the corresponding energy expenditures increased by 17 percent to 4.2 cents/$VA.

The six primary product industries described earlier accounted for 83 percent of the energy consumed by the manufacturing and chemical sector in 1971. On a Btu/$VA basis they were roughly nine times as energy intensive as the remaining smaller industries, which were not represented but are also normally considered as part of this sector. The food-processing industries account for only 6 percent of this total. The other five primary product industries are eleven times as energy intensive as the excluded industries. This means that in 1971 these five industries alone consumed 14 cents of energy/$VA.

It is hoped that the prospect of rising energy costs and uncertainties in future supplies will result in industries rethinking their energy use and undertaking in-house programs to improve energy efficiency. A major study[2] suggests that the difference between the projected and historical rate of decrease in energy consumption per $VA is the result of industry efforts to economize in their use of scarce energy resources. Table 7-1, taken from that study, suggests that, for the six most energy-

TABLE 7-1 Projected Energy Output Coefficients for the Six Energy-Intensive Industries

	1971 Energy Requirements		1980 Energy Requirements		1971–1980 % Decrease (Per year)	1990 Energy Requirements		1971–1990 % Decrease (Per Year)
	10^3 Btu/$VA	10^6 J/$VA	10^3 Btu/$VA	10^6 J/$VA		10^3 Btu/$VA	10^6 J/$VA	
Food and kindred products	31.98	33.74	26.29	27.74	17.8 (2.15)	26.47	27.93	17.2 (1.00)
Paper and allied products	115.80	122.17	94.31	99.50	18.6 (2.26)	81.15	85.61	29.9 (1.85)
Chemicals and allied products	95.70	100.96	84.02	88.64	12.2 (1.46)	72.76	76.76	24.0 (1.43)
Petroleum and coal products	451.00	475.81	373.55	394.10	17.2 (2.07)	370.57	390.95	17.8 (1.03)
Store, clay, and glass products	146.00	154.66	119.55	126.13	18.5 (2.24)	105.90	111.72	27.4 (1.70)
Primary metals industries	212.66	224.36	177.90	187.68	16.3 (1.96)	154.50	163.00	27.3 (1.67)
TOTAL	128.04	135.08	107.49	113.40	16.0 (1.93)	93.32	98.45	27.1 (1.65)

SOURCE: U.S. Department of Commerce, National Technical Information Service PB-248-495. Prepared by Federal Energy Administration, FEA/N-74/536, November 1974.

intensive industries between 1971 and 1980, the energy consumption per $VA is expected to decrease by roughly 1.93 percent/yr. The impact of this projected decrease on the baseline energy consumption growth rate for the entire manufacturing and chemical sector in energy consumption is evident from inspection of Table 7-2.

7-3 INDUSTRY POTENTIALS FOR ENERGY–FUEL USE REDUCTION

According to the U.S. Department of Commerce, Bureau of Census, five major manufacturing categories account for roughly 75 percent of the total industrial fuel purchases. They are:

1. Pulp and paper industry
2. Iron and steel industry
3. Cement industry
4. Chemical industry
5. Petroleum refining

It may be instructive to evaluate the fuel and energy impacts of these industries since they are generally representative of the problems faced by the chemical and manufacturing sector. Our study will help us evaluate the potentials for meaningful fuel substitution and ECMs.

7-4 PULP AND PAPER MANUFACTURE

The U.S. paper, pulp, and related paperboard industry is the fourth largest consumer of fuel and energy among manufacturers. Approximately 40 percent of that energy is derived from its own process wastes, as shown in Table 7-3.

See Fig. 7-3 for an illustrative flow sheet of a typical kraft pulp and paper processing facility. If we break down a paper mill operation by energy flows, we discover some interesting facts. Debarking and chipping account for roughly 1 percent of the total energy consumption; pulping takes 41 percent, and bleaching requires 17 percent. The remaining 41 percent is required by paper forming and drying operations. A typical bleached folding box board, for example, will consume 14,000 Btu of energy for each pound of product (670,500 kJ/kg of product) produced. A pound of product made from primary fiber requires more total energy than a ton of secondary fiber, which is obtained from recycling internal and external scrap.

Paper mills using primary fibers averaged approximately 15,000 Btu/lb (7615 kJ/kg) of product, whereas paper mills employing recycled or secondary fibers

TABLE 7-2 Impact of Improved Energy Efficiency on Manufacturing-Sector Energy Requirements

	1971 Energy Requirements		1971–1980 Projected Growth in Energy, %/yr	1980 Energy Requirements		1971–1990 Projected Growth in Energy, %/yr	1990 Energy Requirements	
	10^{12} Btu	10^{12} kJ		10^{12} Btu	10^{12} kJ		10^{12} Btu	10^{12} kJ
Baseline manufacturing energy consumption	16,085	16,984	5.90	26,945	28,450	5.2	40,646	42,916
Impact on projected energy requirements:								
1. Changes in output mix			−0.61	1,389	1,466	−.61	−2,933	−3,096
2. Efficiency improvements in six energy intensive industries			−1.64	−3,333	−3,519	−1.21	−7,453	−7,869
3. If other manufacturing firms improve efficiency at same rate as six energy intensive industries			−0.39	−740	−781	−.33	−1,808	−1,909
Projected manufacturing energy requirements	16,085	16,984	3.26	21,474	22,673	3.05	28,452	30,042
Reduction below baseline			−2.64	5,471	5,776	−2.15	12,194	12,875
				Equiv. bbl oil/d	Equiv. m³/day		Equiv. bbl oil/d	Equiv. m³/d
				2.50	0.4		5.57	0.89

SOURCE: U.S. Department of Commerce, National Technical Information Service, PB-248-495, p. 10. Prepared by Federal Energy Administration, FEA/N-74/536, November 1974.

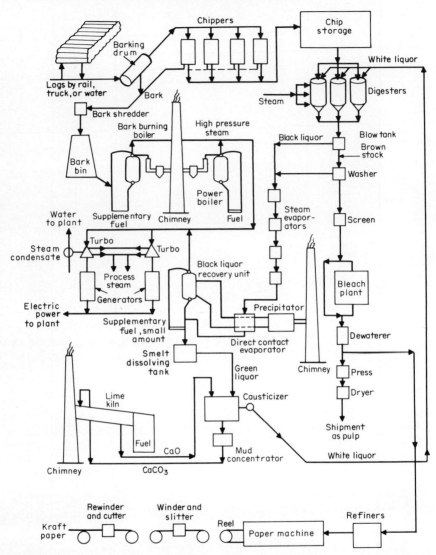

Fig. 7-3 Flow sheet for kraft pulp and paper. ("Intra Industry Capability to Substitute Fuels," Federal Energy Administration, report no. FEA-E1-50034, October 1974.)

TABLE 7-3 Sources of Energy in the Pulp and Paper Industry

Spent pulping liquors	32.8%
Hogged wood and bark	6.9%
Natural gas	20.8%
Distillate fuel oil	2.0%
Residual fuel oil	20.0%
Coal	11.6%
Purchased electricity	4.8%
Other (propane and purchased steam)	1.0%
	100.0%

SOURCE: U.S. Department of Commerce, National Technical Information Service PB-237-605, p. 43. Prepared by Science Communications, Inc., October 1974.

required only 9700 Btu/lb (4646 kJ/kg) of product. The primary fiber mills, however, utilize internal energy sources for 64 percent of their fuel. Thus, they were able to reduce their net consumption to 5700 Btu/lb (2730 kJ/kg). Therefore, the major advantage of recycling is the conservation of our forest resources from which the pulp wood is derived.

In reviewing recent American Paper Institute reports[1] before Congress, additional energy savings opportunities in pulping technology, on paper machines, in the power plant, and effluents included:

1. Enhanced pulp yield per ton of wood chips.
2. Development of oxygen-bleaching technology focused on reducing energy input and waste treatment problems.
3. Increased efficiency in water removal in the press section and in various drying and ventilating techniques.
4. Improved machine control techniques which permit greater paper and paper board yield per ton of pulp.
5. Increased utilization of white water in processing.
6. Use of large power generation units employing improved combustion apparatus, improved heat traps, and better use of automatic controls.
7. Process improvements to reduce losses of chemical, fiber, and other solids.
8. Increased energy yields from waste products.
9. Employing a closed water cycle to permit recapture of low-temperature heat in warm water effluents.

A major paper company[1] reported the following items for its current task list:

Utilities	Maximize corporate hydroelectric generation
	Maximize steam electrical extraction generation
	Maximize use of by-product boiler fuels—bark and waste liquor
	Employ improved boiler efficiencies
	Maximize condensate return
Paper mill	Water recycling—fiber and heat recovery
	Shower water temperature reductions
	Shower water flow reduction
	Yankee hood operational improvement
	Heat recovery economizers
Pulp mill	Heat recovery of pulp drier exhausts
	Heat recovery from evaporator condensate
	Replace low-consistency bleaching with high-consistency bleaching
	Analysis of major electric drives to increase efficiency

Today's paper and pulp industry is making a maximum effort to broaden fuel conservation programs and reduce its dependence on oil and natural gas. However, industry officials have expressed concern over the inadequacy of federal government assistance in overcoming present barriers to the use of coal.

7-5 IRON AND STEEL INDUSTRY

The iron and steel industry uses[1] approximately 3.25 quads (3.43×10^{15} kJ), or 17 percent of the U.S. industrial energy consumption. This corresponds to roughly 5 percent of the total. The various industry energy inputs break down roughly as follows: 68.9 percent for coal, 20.6 percent for natural gas, 6.5 percent for oil, and 4.0 percent for purchased electricity.

In nonintegrated mills, the balance shifts to 7 percent from coal, 6.3 percent from natural gas, 6 percent from oil, and 24 percent from purchased electricity. Iron and steel industry use patterns are materially affected by by-product gases from coke ovens and blast furnaces, as well as by other energy-rich by-products. Depending upon the characteristics of the steel-making processes used and the available raw material quality, fuel use patterns for steel mills generally vary according to the degree of integration employed. As a result, any attempt to shift to a greater use of coal must take into account the interdependency of existing energy recovery cycles. A typical distribution of energy use (segregated by process) for the iron and steel nonintegrated industry is given in Table 7-4.

Bear in mind that experience varies somewhat from company to company and from mill to mill. There are also frequent differences in methods of accounting for by-product energy credits.

The primary use of coal in steel making is for coke ovens. Fuel values are recovered as coke oven gas, breeze, tars, creosote, and salable aromatic chemicals

TABLE 7-4 Energy Usage for Iron- and Steel-Making Operations

	Industry Average, %	An Integrated Mill, %
Raw materials preparation	5.6	4.8
Coke making	17.9	11.1
Iron making	35.0	57.9
Steel making	13.2	5.2
Primary Rolling	6.3	17.6
Finishing	11.4	3.4
Other	10.2	
	100.0	100.0

SOURCE: U.S. Department of Commerce, National Technical Information Service PB-237-605, p. 53. Prepared by Science Communications, Inc., October 1974.

such as benzene, toluene, and xylene. In a typical operation, fuels used directly in the various iron- and steel-making processes are distributed as follows:

Purchased natural gas	29 percent
Purchased oil	14 percent
Coke oven gas	38 percent
Blast furnace gas	9 percent
Tar and creosote	6 percent
Reclaimed oil, coke breeze, and tar	3 percent

It may be instructive to examine source energy flows in the various industry key steps. The steps are:

1. Preparation of raw materials
2. Manufacture of coke
3. Manufacture of iron
4. Manufacture of steel
5. Rolling, reheating, and finishing

Initial operations occurring either at the mine or at the iron works involve, respectively, the pelletizing of low-grade ore or the sintering of slag and ore. The sintering process is highly energy-intensive; it consumes an average of 2.2×10^6 Btu/ton (2.6 GJ/mt) of processed ore. Most of the energy used, however, is supplied from the combustion of by-product coke breeze.

Purchased fuel is employed only for initial ignition of the bed. It is anticipated that, by 1990, pelletizing will generally replace sintering as the primary preparation method for changing the ore to the blast furnace. This is a result of longer distances between mine and mill and the decreasing overall quality of available domestic iron ore. Although pelletizing results in 63.6 percent savings in energy per ton of ore processed, it requires a fuel mix of roughly 75 percent natural gas and 25 percent oil. The substitution of coal in pelletizing or in electric arc furnaces will require further development. At present only the SL/RN process[1] can be used to fire a pelletizing furnace to the required operating temperatures on the order of 2000 to 2400°F (1100 to 1320°C).

In recent years, the manufacture of coke has become highly energy-efficient, with a reported overall 90 percent energy recapture when all by-product credits are included. Although coke ovens are externally heated, fuel consumption amounts to approximately 5 percent of the input energy which is provided by combustion of coke oven gas and by-product tars. Presently coke is quenched with water as it leaves the oven. However, environmental pressures, principally caused by resulting air pollution problems, point to greater future use of dry quenching. Dry quenching, now prevalent in Europe, results in energy savings of 20 percent of the net energy consumption levels for coking, or roughly 1×10^6 Btu/ton (1.16 GJ/t) of coke produced.

Due to high capitalization costs, however, it is currently estimated that average fuel prices will have to rise to $3.00 to $3.50/$10^6$ Btu ($2.95 to $3.32/GJ) before dry quenching can be considered cost-effective in U.S. operations.

The manufacture of iron takes place exclusively within a blast furnace. The process is regarded as the most energy-intensive of the iron- and steel-making processes. Even after we account for the credit available from blast furnace gases, iron making consumes an estimated 13.5×10^6 Btu/ton (15.66 GJ/t) of pig iron produced. Although the principal energy source is derived from coke charge, a decline in the availability of coking coals makes continued reliance questionable.

Accompanying this process is an injection of hydrocarbons (generally fed through the combustion-air supply ports) at an estimated energy input of 1.5×10^6 Btu/ton (1.74 GJ/t) of pig iron produced. This results in a 10 percent increase in furnace capacity. It is estimated that hydrocarbon injection rates will continue to rise in the future. Some companies are experimenting with injected pulverized coal, in lieu of hydrocarbons, without any reported sacrifice in furnace performance or capacity. However, increased substitution of coal is limited by the fact that many of the presently injected materials represent coke oven by-products.

Another process which has received considerable attention abroad during recent years is the direct reduction of iron ore involving substitution of hydrogen or hydrocarbons entirely for coke. However, because of the unique supply and price relationships of coal and natural gas likely to continue in the United States for some time, its domestic impact is believed to be negligible.

Auxiliary gas or steam for driven blowers required for inducing draft in blast furnaces presently consumes approximately 2.5×10^6 Btu/ton (2.9 GJ/t) pig iron. This energy generally comes from by-product gas or waste steam.

7-6 STEEL MANUFACTURE

Steel making can be accomplished by one of the three conventional processes:

1. Open hearth furnace
2. Basic oxygen furnace
3. Electric furnace

These processes have widely differing energy demand profiles. A typical open hearth furnace requires an energy input of 3.0×10^6 Btu/ton ingot (3.48 GJ/t ingot). Depending upon operating conditions, this value may range ± 17 percent. Most of the energy is supplied by by-product streams, with the balance provided by combustion of oil and/or natural gas. Due to the regenerative nature of open hearth operations on the heating cycle, it would be rather costly to employ conventional coal substitution methods (such as the use of pulverized coal). The open hearth process, however, is being phased out. Therefore, further process improvements are considered unlikely.

The basic oxygen furnace is expected to account for more than 70 percent of U.S. steel production by 1990. The basic oxygen furnace is inherently more efficient than the open hearth furnace; it represents an approximate savings of 90 percent per ingot ton produced. Furthermore, improvements in overall energy requirements can be made possible by more efficient heat recovery apparatus. But here again it is estimated that energy prices must rise to a level just above $3.00/10^6$ Btu ($2.84/GJ) to justify the additional capital investment necessary.

The major energy cost associated with operation of the basic oxygen furnace is indirect. It involves the off-site production costs of oxygen necessary to fuel the furnace. Present cryogenic processes used in the manufacture of oxygen (e.g., air-reduction methods) utilize purchased electric power. Consequently, opportunities for a reduction in gas and oil fuel use require the conversion of existing fossil fuel generation stations to coal.

Finally, the use of the electric furnace process in the manufacture of steel is expected to increase in the years ahead. Although the electric furnace process requires the lowest energy input per ingot ton of all three processes, it is entirely in the form of electricity. It is, therefore, subject to coal substitution only at the serving utility power generating stations.

The use of the electric furnace has not had a significant impact on steel manufacture to date. One reason is that the electric furnace process is simply not competitive with the open hearth or basic oxygen furnace process where molten iron is available for charging. However, when scrap steel is available for charging, the process is considered effective. Increasing the availability of scrap steel thus enhances opportunities to reduce the energy intensiveness of steel manufacture, while permitting a shift to coal.

Rolling, reheating, and finishing operations in the manufacture of various steel products account for a substantial proportion of the energy consumed. These energy flows are generally distributed among a variety of heating and forming operations. In such operations, usually performed at the steel mill, approximately 50 percent of the total energy requirements are provided through the combustion of coke breeze and by-product gas. The principal energy source for metal-forming operations is provided by electricity or steam-driven machinery. Purchased fuel, such as natural gas, is principally used as a standby fuel or for pilot ignition. Fuels for soaking pits and reheat furnaces must be relatively ash-free to avoid troublesome operational problems. Therefore, immediate coal substitution at this end of steel manufacture is considered unlikely. A long-term possibility for coal fuel substitution lies with the development of commercially available coal gasification or liquefaction products.

7-7 CEMENT MANUFACTURE

In Chap. 1 we dealt with the energy-intensive nature of cement as a construction material. The cement industry has been accused of representing an excessive energy cost per dollar of raw material. In our earlier discussion, however, we pointed out that on a per ton basis, many other construction materials, including steel, account for larger consumptions of energy. The cement industry consumes approximately 1.5 percent of the total industrial energy, or roughly 0.8 percent of the total U.S. energy use. In a recent year, the cement industry comprised some 51 separate companies operating 170 plants in the United States. They had a total capacity of approximately 85 million tons/yr (99t/yr).

Cement was a key constituent estimated at greater than 90 percent of all construction. Yet it accounted for only 1.4 percent of the total construction expenses of roughly $136 billion. If we examine the operations side of cement manufacture, we find the kilns consume 80 percent of the total fuel required in the manufacture of cement. It is here that mixtures of limestone and clay are calcined to form a cement clinker. Another 13 percent of the total fuel consumption is used for grinding materials. The balance is required for material handling, product losses, etc.

The cement industry has become more energy-conscious in recent years. It recognizes that energy costs represent a significant portion of total manufacturing costs. Recent short-term ECMs, which have resulted in a reduction of fuel consumption of about 15 percent on a per ton product basis, include:

1. Installation of chain sections to maximize fuel efficiency
2. Various kiln alternatives and general upgrading
3. Improved quarrying and materials-handling operations
4. Adjustments of raw material balance

Industry sources suggest that all cement-producing plants can be converted to coal firing. However, the estimated cost for kiln conversion in a typical plant ranged from $2.5 to $3.5 million. Such a conversion required one year from the date of order to the date of equipment delivery. The financial investment represented approximately 5 percent of the entire original cost of the plant. Limited availability of capital for conversion or expansion of present capacity is a growing concern for operators. The problems of limited coal availability and transportation logistics must also be considered.

7-8 CHEMICAL AND ALLIED INDUSTRIES

Chemical and allied industries represent the largest single industrial energy consumer. Their projected consumptions for five representative years from 1968 through 1990 are listed in Table 7-5. These industries fall into five generalized subindustry groupings:

1. Industrial gases (e.g., acetylene)
2. Chlorine-caustic industry
3. Industrial organic chemicals (e.g., ethylene)
4. Inorganic pigments (e.g., zinc oxide)
5. Industrial inorganic chemicals (e.g., ammonia)

Approximately 1000 different primary products are produced by the above industry groupings. Because of gross variations in fuel requirements, process chemistry, feedstock, energy recovery potentials, etc., our attention here will focus on only three basic chemicals whose manufacture accounts for more than 50 percent of the total chemical and allied industries energy usage. They are:

1. Ammonia
2. Ethylene
3. Chlorine

Ammonia is the primary nitrogen fertilizer and intermediate for the manufacture of numerous other widely used chemicals, including other nitrogen fertilizer products. Approximately 35×10^6 Btu (36.9 kJ) is consumed for each ton of anhydrous ammonia produced. Most of this energy is in the form of heat. Compression, by means of steam- or electrically driven equipment, accounts for the balance. Steam turbines are more generally used for compression, since process ammonia reformers already produce by-product steam during the cooling of exit gases. It is difficult to generalize about opportunities for the substitution of coal for oil or gas, since they vary directly with economic and technological constraints for each individual plant.

TABLE 7-5 Projected Net Energy Consumption in Chemical and Allied Products Manufacturing, 10^{12} Btu and 10^{12} kJ

	1967	1977	1980	1985	1990
Basic inorganics*	1007 1063	1404 1482	1542 1628	2058 2173	2234 2359
Basic organics†	917 968	1741 1838	1831 1933	2237 2362	2783 2938
Synthetics‡	379 400	729 769	790 834	1030 1087	1340 1415
Finished products§	295 311	554 585	659 696	850 897	1124 1187
TOTAL	2598 2743	4428 4675	4822 5091	6175 6520	7841 7899

*Inorganic acids, alkalies, salts, and gases (also enriched uranium).
†Organic solvents, acids, salts, and intermediate petrochemicals.
‡Plastics, synthetic fibers, and synthetic rubbers.
§Cosmetics, soaps, drugs, paints, and explosives.

SOURCE: U.S. Department of Commerce, National Technical Information Service PB-248-495, p. 3-2. Prepared by Federal Energy Administration, FEA/N-74/536, November 1974.

Industrial Sector Conservation Opportunities 173

In the case of ethylene, substitutions between coal, oil, and gas as energy sources are complicated by the fact that ethylene manufacture is one of many feedstreams within an integrated petroleum refining or petrochemical complex. Changing over from one fuel to another affects all or other portions of the complex operations. The three principal manufacturing methods of ethylene include:

1. Cracking of naphtha to produce intermediates ethylene and propylene.
2. Cracking ethane propane or *n*-butane fractions (obtained from natural gas).
3. Cracking crude oil directly (i.e., Union Carbide's process anticipated to be in commercial use soon).

The United States produced approximately 8 million tons (9.31 t) of chlorine, and the entire chlorine-caustic industry consumes in excess of 0.3 quad (0.3×10^{15} kJ) of energy annually.

The two principal processes employed to manufacture caustic soda, in addition to chlorine and hydrogen gases, include:

1. The diaphragm cell which produces a low-concentration caustic soda solution by electrolyzing prime oil.
2. The mercury cell which produces a higher-concentration caustic soda solution by electrolyzing brine.

There has been a pronounced shift away from the use of the less energy-intensive mercury cell process in recent years. Environmental concerns brought on by mercury (waste) poisoning episodes in Japan and the Great Lakes have retarded commercial development. Work is now proceeding on mercury cell plant effluents to reduce mercury trace levels. At the same time, efforts are being made to improve the energy efficiency of the diaphragm cell through:

1. Development of a new dimensionally stable anode which is more efficient than presently used carbon anodes.
2. Improvements to cathode and electrolytic cell geometry resulting in higher cell efficiencies.
3. Staging of evaporators used in concentrating caustic soda solution.
4. Greater use of more efficient combined on-site steam-power generating units.
5. Development of new synthetic diaphragms to eliminate the need for caustic soda solution concentration (anticipated by 1980–1985).

From our brief review of the vast chemical and allied industries, we can conclude that opportunities for meaningful energy savings fall into two general categories:

174 Energy Conservation in Buildings and Industrial Plants

1. Those processes in which a portion of the material stream serves a fuel function and another portion serves a process function.

2. Those processes where the heat of reaction (exothermic) of heat (endothermic) required for reaction produces a waste heat flow that can be utilized within the process as an energy source.

Significant steps have been taken to improve energy conservation in the chemical process industries. A report from the Manufacturing Chemists Association near the end of 1978 indicated that chemical industries were operating with 16 percent less fuel consumption per unit of output as compared with prevailing rates in 1972. Many industry officials feel that, had it not been for the need to comply with new workplace and environmental regulations, the savings in energy use would have been even greater.

7-9 DIRECT CONVERSION OF CHEMICAL ENERGY

The fuel cell is a device which converts chemical energy directly into electrical energy and waste products. Reactants—a suitable fuel and an oxidant—are supplied to the device, and the products of the reaction are discharged continuously. See Fig. 7-4.

Fuel cells can operate on a variety of hydrocarbons, ranging from natural gas to synthetic, low-Btu gases derived from coal or the pyrolysis of municipal solid waste. Such readily available fuels can be fed with steam, which can be recycled from the fuel cell stack in situations where fuel cells are piled one on top of the other. As DC electricity is supplied to processes served by it, exhaust, containing harmless amounts of carbon dioxide, nitrogen, and water vapor condensed from the steam, is discharged from the stack into the atmosphere.

The theory of operation will not be discussed here, since a variety of different

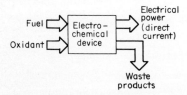

Fig. 7-4 The basic fuel cell. (*Energy Fact Book, 1977*, Tetra Tech, Inc., Arlington, Virginia, April 1977, AD/A-038 802, p. XIII-1b.)

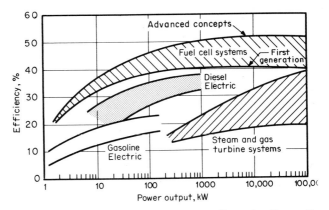

Fig. 7-5 Comparison of power steam efficiencies (*Energy Fact Book, 1977*, Tetra Tech, Inc., Arlington, Virginia, April 1977, AD/A-038 802, p. XIII-8.)

design approaches have been demonstrated to date. The more common types now in use or under development include:

1. Direct-type, alkaline electrolyte
2. Bacon cell
3. Union Carbide cell
4. Ion-exchange membrane cell
5. High-temperature cell
6. Redox cell
7. Hydrazine fuel cell

Much research is being directed toward making a more practical, lower-cost device suitable for use in various high-power density applications.[3]

In terms of efficiency, the fuel cell stacks up well against other types of power generators. This is evident in Fig. 7-5. The curve labeled "advanced concepts" refers to a class of high-temperature cells which also employ hydrocarbon fuel. After being converted into hydrogen and carbon monoxide, carbonates are employed as a weak alkali. High-temperature cells feature current densities at approximately 0.7 V, which vary from 20 to 100 A/ft^2.

In the final analysis, economics becomes the determining factor. One method of comparing energy conversion systems is to combine fixed (capital) costs and operating (proportional) costs to develop total cost that relates the system load factor to the relative cost per kilowatthour. This method is illustrated in Fig. 7-6. Note that if we had a design requirement for a 53-kW load, in terms of total cost it would be more

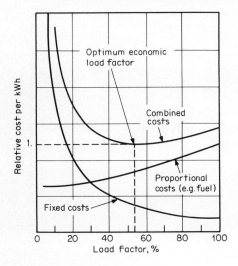

Fig. 7-6 Cost variations as a function of load factor. (*Energy Fact Book, 1977*, Tetra Tech, Inc., Arlington, Virginia, April 1977.)

advantageous to consider a 100-kW power plant. Such a plant would have adequate reserve and the capacity for future expansion. It could also accommodate a 40 percent reduction in capacity without any penalty in performance, as in the relative cost per kilowatthour.

7-10 PETROLEUM-REFINING INDUSTRY

Petroleum refining is both a major consumer of energy and provider of basic energy fuels. In the late 1960s and early 1970s, prevailing prices for oil and gas were not sufficient to attract the kind of capital needed for developing prospects to replace fuel resources consumed. The industry began to use up its reserves. The liquidation of reserves was interrupted only in 1970, by the Prudhoe Bay discovery, and then accelerated again.

Higher prices have dramatically broadened opportunities for domestic exploration; the number of prospects that offer a reasonable rate of return has increased geometrically. Moreover, oil and gas prices have risen sufficiently to keep the loss of the percentage depletion allowance from significantly lessening the economic attraction of exploring for oil and gas.

Energy consumed in petroleum-refining operations can be classified under three major generalized headings:

1. Production of process steam, which accounts for 25 percent of total refining energy inputs.
2. Direct process heat, which accounts for 60 percent of the total refinery inputs.

3. By-product gas streams, catalytic coke, fuel oil products, propane and butane products, purchased natural gas and purchased (or generated on-site) electricity, which account for the remaining 15 percent of total refinery energy inputs.

Energy consumption can be materially reduced by improving maintenance practices, increasing heat recovery from process flows, and incorporating modifications to refinery operations. Conservatively, a 10 percent reduction in the energy consumption of a typical major refinery is believed possible by various combinations of the above proposed measures. But full implementation could take up to an estimated four years.[1]

In refining operations, the opportunities for coal substitution appear rather limited. They are generally confined to fuel substitution for on-site steam generation facilities where coal availability, physical space for equipment and storage, and environmental constraints are not limiting. Coal price and availability must also be carefully considered. Other factors include additional costs associated with new coal-fired boilers, coal-handling facilities, stack gas cleanup equipment, and return on investment.

7-11 EVAPORATION AND DISTILLATION PROCESSES

A key industrial opportunity for significant advances in conserving energy lies in the selection and optimization of chemical process operations.[4,5] Because of anticipated industrial output growth and because of the need for new and larger sources of environmentally acceptable energy sources, the chemical industry must consider modifications in process design which permit a higher utilization of such energy sources.

To justify the additional capital expenditures required, we must establish a proper relationship between capital and operating costs and the time value of money. Additional factors include project life, taxes, anticipated maintenance and start-up costs, depreciation, working capital, and insurance. In the final analysis, the result can often be expressed as some ratio of capital and operating cost, as illustrated in Fig. 7-7 for optimization[4] of a heat exchanger. A common overestimate in this kind of analysis is the anticipated availability of the energy source selected. Also bear in mind that such a plant does not usually operate at its design point at all times; we must consider a likely operating range and the weighting of anticipated annual loading.

We can, for example, compute the capital equivalent of incremented power, employing time-weighted discounted cash flow in methods expressed as dollars per kilowatt, dollars per cubic foot (or cubic meter), or 1000 lb (454 kJ) of steam, at a given set of operating conditions. Once these financial data are developed, we establish the system performance curve. Its shape can prove a key factor in optimum

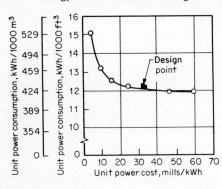

Fig. 7-7 Heat exchanger optimization. Increment between adjacent points represents equal changes in investment. (R. L. Shaner, "Energy Scarcity: A Process Design Incentive," *Chemical Engineering Progress*, May 1978, p. 47.)

equipment selection. In Fig. 7-7, notice that beyond the knee of the curve there is less economic incentive to further reduce energy consumption. This is because, while pressure drop is reduced (thereby requiring less power to drive the system) by increasing cross-sectional flow area, we soon reach a point where further additives in the cross-sectional area do not justify the additional capital expenditure. Understanding the so-called "knee" curve relationship is essential in evaluating process energy conservation opportunities. Recent studies[4] suggest that more than 75 percent of the heat transferred in industrial processing units occurs at 350°F (177°C) or below.

Considering the design problems on a system basis (i.e., interchanging high-temperature heat sources with heat consumers elsewhere in a given process, or by heat pumps) provides the process designer with important conservation opportunities.

In analyzing such opportunities, we must apply the record low. The record low analysis permits us to eliminate all unnecessary irreversibilities resulting from excessive temperature differences, pressure drops, equipment inefficiency, and heat losses.

To illustrate this principle, let us consider the operation of a conventional C_3 splitter employed by chemical process industries in the separation of propylene from propane, as illustrated in Fig. 7-8. Normally, this column is operated at 300 psig (2067 kPa) so that cooling tower water can be utilized for condensation of the overhead propylene product.

If the tower operating pressure is reduced to 125 psig (861 kPa), improved separation is made possible for these close-boiling constituents by employing a heat pump cycle, as illustrated in Fig. 7-9. Reducing the number of trays, the reflux ratios, the column pressure drop, and required vessel wall thicknesses results in further savings. Improving the heat pump compressor effectiveness, as illustrated in Fig. 7-10, brings about a further reduction in energy consumption.

Refer to Table 7-6 for an economic comparison of process design options. Notice that significant amounts of energy and capital can be saved by improved heat pump operations. Distillation columns, operating at varying temperatures, often present the opportunity to exchange (or heat pump) any heat surplus from one tray to another

Industrial Sector Conservation Opportunities   179

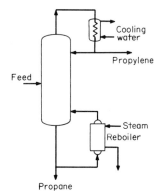

Fig. 7-8 Conventional C₃ splitter. (R. L. Shaner, "Energy Scarcity: A Process Design Incentive," *Chemical Engineering Progress*, May 1978, p. 48.)

requiring heat. Chemical reactors, though somewhat less flexible in operating temperature than distillation columns, can also be designed with a widened temperature availability range to improve energy conservation potentials.

One of the most promising techniques involves the use of heat pumps in distillation systems. Although such methods have long been used in chemical processing, the need to achieve further operating savings has prompted a recent upsurge in interest.[6] This stems principally from the fact that, in conventional distillation systems, energy is used on a once-through basis. It normally enters from some high-temperature heat source and exits to some low-temperature heat sink. The cost of separation is usually quite high, since the gross energy is totally degraded from a combustion process [in excess of 3500°F (1927°C) in a conventional oil- or gas-fired generation system] to ambient as established exhaust-air temperature leaving spray type of natural draft cooling tower. If we compare a mechanically driven heat pump with a conventional distillation system with an optional interboiler, we find a significant difference. Instead of using a high-temperature heat source, the heat pump

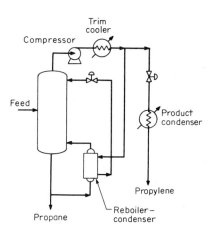

Fig. 7-9 Heat-pumped C₃ splitter. (R. L. Shaner, "Energy Scarcity: A Process Design Incentive," *Chemical Engineering Progress*, May 1978.)

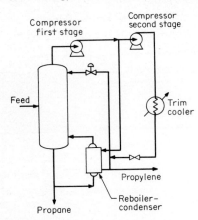

Fig. 7-10 High efficiency C₃ splitter. (R. L. Shaner, "Energy Scarcity: A Process Design Incentive," *Chemical Engineering Progress*, May 1978.)

takes the energy from the condenser and uses only that work necessary to elevate it to a temperature level to be column-reboiled.

Another technique showing promise involves the use of Peltier-effect diffusion stills (PEDS).[7] PEDS apply heat pump principles for close-boiling fluids similar to those frequently encountered in the chemical process industries. Through unique geometry and design, they permit multiple reuse of initial energy sources.

During the late sixties and early seventies, a variety of technical problems arose concerning various thermoelectric device junctions, particularly those employing shoe and strap arrays. As a result, earlier broad-based financial support for development of large-scale thermoelectric devices and systems slackened just as major breakthroughs in promising materials, fabrication, and assembly methods seemed imminent. Subsequent studies revealed that the technical problems could be overcome by employing one or more of the following methods: broader area contacts, namely, direct fusion of the thermoelectric material to the metal strap by heating above the melting point or eutectic temperature; solid-state diffusion bonding, or direct soldering of the joint; and wetting of both the strap and the thermoelectric element. The success of new thermoelectric power generators and the U.S. Navy's nuclear submarine life-safety systems employing thermoelectric air conditioning systems[7] further confirmed the reliability of current state-of-the-art junction methods.

The PEDS concept, then, was originally conceived several years ago, when the availability of low-cost fuels made it difficult to promote commercially despite PEDS' demonstrated high thermodynamic operating efficiency. Although a proof-of-principle apparatus was described in the literature,[7] additional development remains in the refining of hardware and in appropriate thin-film manufacturing techniques.

PEDS are characterized by a central apparatus comprising concentric series of sections which are closely spaced to define circular diffusion spaces between them. See Fig. 7-11. Within such sections, a series of thermoelectric (p/n type) elements are positioned. They operate as heat pumps by providing heated and cooled surfaces,

TABLE 7-6 Economic Comparison of Process Design Options

	Case 1: Conventional System	Case 2: Heat Pump Conventional Equipment	Case 3: Heat Pump High-efficiency Equipment
	Equipment Costs, Installed, thousands of $		
Columns	5400	2022	1522
Reflux pump	25		
Heat exchangers	1172	748	865
Compressor and drive (steam turbine)	—	2346	1313
Totals	6597	5116	3700
	Utility Costs, thousands of $/yr		
Steam	1488*	810†	453†
Cooling water‡	258	53	28
TOTALS	1746	863	481

*High-pressure steam (turbine), $2.50/1000 lb.
†Low-pressure steam (reboiler), $1.75/100 lb.
‡Cooling water, $0.05/1000 gal.
SOURCE: R. L. Shaner, "Energy Scarcity: A Process Design Incentive," *Chem. Eng. Prog.* (May 1978).

respectively, on opposite sides of the section. These elements face each other across these spaces to evaporate (from hotter surfaces) and condense (on colder, juxtaposed surfaces) a feed fluid flowing by gravity over heated surfaces. Means are provided to collect the condensate from the colder surfaces.

The primary difference between diffusion and distillation involves the mass transport potential, which has a profound effect on the economic feasibility of the proposed methods. Contrary to flash distillation, the heat of vaporization is supplied uniformly at the finned evaporating surfaces of the annular thermoelectric sections. This is illustrated in Fig. 7-12.

In Fig. 7-12, when vaporization occurs at the evaporating surface a with fluid at temperature T_e, water molecules leave the evaporating liquid, diffuse through a vapor mixture between closely spaced finned surfaces, and condense on the adjacent surface b where condensate is formed. The rate of heat transfer of the evaporating fins is a function of film thickness, flow characterization, etc. Once the entire heat transfer surface in the evaporator is wetted with feed, the evaporation rate increases with the feed rate.[8]

By maintaining the condensing surface b at a temperature T_c lower than the adjacent evaporating surface a, partial pressure and temperature gradients are established which maintain the mass transfer process. At the close spacing established

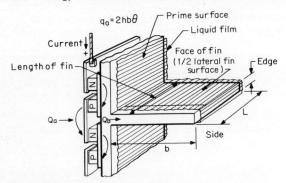

Fig. 7-11 Typical fin and film profile. (Milton Meckler, "Peltier Effect Heat/Mass Transfer Systems," address before the Fifth International Heat Transfer Conference, Tokyo, September 3 to 7, 1974.)

between adjacent condensing and evaporating surfaces and the small differentials maintained between them, the condensing liquid film is diffusion-controlled with negligible convection and radiation effects.

The rates of mass transfer are higher than in a comparable distillation apparatus operating between the same condensing and evaporating temperatures, because the rate of diffusion of the water vapor across the gap X, separating juxtaposed evaporating and condensing fins, is a function of the relative partial pressures of the liquid surfaces a and b requiring separation.

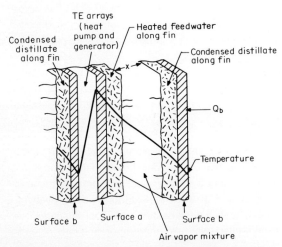

Fig. 7-12 Basic principles of multistage diffusion. (Milton Meckler, "Peltier Effect Heat/Mass Transfer Systems," address before the Fifth International Heat Transfer Conference, Tokyo, September 3 to 7, 1974.)

Some representative applications of a PEDS diffusional process involve desalination, multicomponent distillation, and reactor design. Such applications employ available waste heat steams, direct solar energy conversion means, etc., to significantly reduce operating costs by reuse of thermal loads generated within the very processes themselves. A PEDS apparatus performing binary (or multicomponent) separations is diagrammed in Fig. 7-13. This apparatus geometry results in an extremely compact heat pump array, modular in construction, capable of operating efficiently over a broad range of conditions, and comprising few moving parts. Such an apparatus requires minimum attention in operation and maintenance.

The types and characteristics of semiconductor materials used in fabricating thermoelectric devices are very similar to those used in manufacturing photovoltaic devices. Furthermore, since PEDS can utilize DC electric power, it would seem to couple well with fuel cell technology. As DC power is produced and supplied to PEDS, exhaust from the stack discharged into the atmosphere would include only harmless amounts of carbon dioxide, nitrogen, and water vapor condensed from the steam. Fuel cells, therefore, represent a natural match with PEDS, and the costly inverter step can be eliminated. A PEDS and fuel cell system would require only two components: the fuel processor and a fuel cell stack, to produce DC power from fuel cell electrochemistry. A further advantage of PEDS is that, since fuel cells operate on a variety of hydrocarbons, ranging from natural gas to low- (synthetic) Btu gases derived from coal or pyrolysis of municipal solid waste or biomass, further strains on

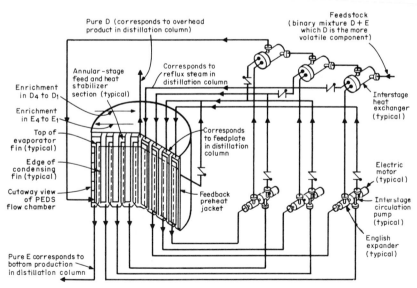

Fig. 7-13 PEDS apparatus performing binary (or multicomponent) separations. (Milton Meckler, "Thermoelectric Diffusion Chemical Systems," New Horizons in Materials and Processing seminar, 18th National Sampe Symposium and Exhibition, vol. 16, Los Angeles, 1973.)

184 Energy Conservation in Buildings and Industrial Plants

electrical generation capacity are eliminated. This permits a wide range of versatility in planning chemical and refinery processing operations.

To date, thermoelectric devices have been employed in a number of smaller power applications.[15,16] PEDS offer their greatest promise, however, in configurations employing electric heat pumps for larger-scale industrial processing. Recognizing that more than three stages are generally necessary for separations employing either PEDS or distillation towers equipped with mechanical heat pumps, the potential energy savings to the petroleum industry alone could substantially reduce the 95 percent energy losses inherent in conventional distillation columns. Even a 10 percent decrease in the amount of energy used for distillation would conserve the equivalent of 100,000 barrels of oil a day. Clearly, there is a need to press ahead with PEDS in refinery and chemical plants. We can no longer afford to postpone the development of the PEDS technology.

Two vapor recompression techniques which have aroused recent interest are the mechanical and the thermorecompression methods, both illustrated in Fig. 7-14. Notice that a steam jet eductor (booster) is employed to recompress a portion of the overhead vapor. In effect, this raises its temperature and pressure to a level where it can deliver its heat of vaporization usefully to the shell-side evaporator section. Energy savings depend upon the differences of steam flow rate (with and without

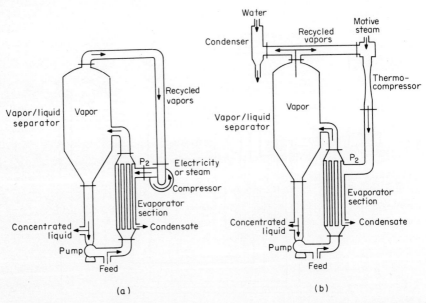

Fig. 7-14 Vapor recompression methods: (a) mechanical recompression and (b) thermal recompression. P_2 = absolute pressure of steam chest. P_1 = absolute pressure of vapor leaving the evaporator. Compression ratio = P_2/P_1. ("Upgrading Existing Evaporators to Reduce Energy Consumption," *Energy Conservation*, 1977, COO/2870-2.)

steam jet), the inlet (motive) steam pressure, and compression ratio (representing the ratio of the absolute pressure of evaporator- (steam-) condensing side to that of the vapor leaving the evaporator body.

As steam costs rise above their current level of approximately $3 to $4/100 lb, ($7 to $9/1000 kg), operating costs attributable to evaporator heat loads can be expected to rise dramatically. Any process which can more efficiently provide multistage evaporation, combined with the equivalent of thermal recompression by heat pumping (or recycling the latent heat of evaporation), will give us an opportunity to make major savings.

The most common technique for separating mixtures of chemical species is distillation. Inefficiencies have long plagued the distillation process. In many instances, more than 95 percent of the energy added to the distillation column at the reboiler is removed by condensers atop the column. Only 5 percent of the energy is not wasted. Economic and operability requirements generally dictate that heat be applied only at the bottom of the tower and withdrawn only at the top. Where energy is inexpensive, it is not usually cost-effective to correct these thermodynamic inefficiencies.

In multistage distillation, heat may be removed at a number of locations in the column. In practice, this concept can rarely be applied at each equilibrium stage, even though it would be theoretically possible if finite quantities of heat were added to every stripping stage and finite quantities of heat were withdrawn from every rectification stage.[9]

Multiple-reboiling systems have the economic advantage of being able to utilize multiple levels of heat. The heat load in a single-reboiler system must have a high temperature because it can only be applied at the base of the column. Because heat energy is generally a function of departure from ambient temperature, this single-input method of reboiling a distillation system is the most costly.

If this same energy is divided up and introduced to the column at several points, the temperature levels of the energy can be lower. Regarding the energy inserted between the bottom tray and the feed tray, the temperature can be lowered progressively as the feed tray is approached. The tower liquid at each point must have a temperature lower than that of the energy source. Only the economic quantity of the heat transfer area in the intermediate reboiler determines how much higher than the tower liquid the energy temperature must be. This concept also applies to the rectification section. Here, heat may be removed at several locations, using successively warmer heat sinks as the feed tray draws near. This is inherently more economical than the use of a single energy sink with a low temperature.

7-12 FORECASTING INDUSTRIAL SECTOR ENERGY CONSUMPTION

The forecast for the industrial sector, extrapolated from preliminary 1974 reports and projected to the year 2000, is shown in Fig. 7-15 and Table 7-7. On a net basis, the

TABLE 7-7 Consumption of Energy in the Industrial Sector, from 1974 Preliminary Reports and Projected to the Year 2000

	1974		1980		1985		2000	
Coal:								
Fuel uses:								
10^6 tons (10^6 t)*	150.7	(136.7)	177	(160)	180	(163)	210	(190)
10^{12} Btu (10^9 kJ)	4,093.3	(4,318.4)	4,600	(4,853)	4,680	(4,937)	5,460	(5,760)
Nonfuel uses:								
10^6 tons (10^6 t)	4.2	(3.8)	8	(77)	10	(9)	18	(16)
10^{12} Btu (10^9 kJ)	114.7	(121.0)	200	(211)	250	(264)	450	(475)
Total coal:								
10^6 tons (10^6 t)	154.9	(139.7)	185	(168)	190	(172)	228	(207)
10^{12} Btu (10^9 kJ)	4,208	(4,439)	4,800	(5,064)	4,930	(5,201)	5,910	(6,235)
Percent of total energy consumption	17.5		18.5		17.1		13.6	
Liquid hydrocarbons:								
Fuel uses:								
10^6 barrels (10^9 L)	637.3	(101.3)	734	(117)	829	(132)	1,023	(163)
10^{12} Btu (10^9 kJ)	3,760	(3,967)	4,380	(4,621)	4,950	(5,222)	6,200	(6,541)
Nonfuel uses:								
10^6 barrels (10^9 L)	499.8	(79.3)	624	(99)	710	(113)	1,080	(172)
10^{12} Btu (10^9 kJ)	2,284	(2,410)	3,120	(3,292)	3,550	(3,745)	5,400	(5,697)
Total liquid hydrocarbons:								
10^6 barrels (10^9 L)	1,137	(180)	1,358	(216)	1,539	(245)	2,103	(334)
10^{12} Btu (10^9 kJ)	6,044	(6,376)	7,500	(7,912)	8,500	(8,967)	11,600	(12,238)
Percent of total energy consumption	25.1		29.0		29.5		26.7	

Gaseous fuels:									
Fuel uses:									
10^9 ft³ (10^9 m³)	9,299.3	(263.2)	8,963	(254)	8,695	(246)	10,377	(294)	
10^{12} Btu (10^9 kJ)	9,494.6	(10,016.8)	9,240	(9,478)	8,920	(9,411)	10,360	(10,930)	
Nonfuel uses:									
10^9 ft³ (10^9 m³)	1,600.7	(45.3)	737	(21)	795	(22)	873	(25)	
10^{12} Btu (10^9 kJ)	1,634.3	(1,724.2)	760	(808)	820	(865)	900	(949)	
Total gaseous fuels:									
10^9 cubic ft³ (10^9 m³)	10,900	(308)	9,700	275	9,490	(269)	11,250	(318)	
10^{12} Btu (10^9 kJ)	11,129	11,741	10,000	10,550	9,740	(10,276)	11,260	(11,879)	
Percent of total energy consumption	46.2		38.6		33.8		25.9		
Purchased electricity:†									
10^9 kWh	617		1,057		1,647		4,302		
10^{12} Btu (10^9 kJ)	2,699	(2,847)	3,600	(3,798)	5,620	(5,929)	14,680	(15,487)	
Percent of total energy consumption	11.2		13.9		19.6		33.8		
Total sector energy inputs,									
10^{12} Btu (10^9 kJ)	24,080	(25,404)	25,900	(27,324)	28,790	(30,373)	43,450	(45,840)	

*Ton = short ton (2000 lb); t = metric ton (1000 kg).
†Includes industrial hydroelectric power.

SOURCE: L. J. William. "Preliminary Forecast of Energy Consumption Through 1985," Electric Power Research Institute, Palo Alto, Calif., March 1976.

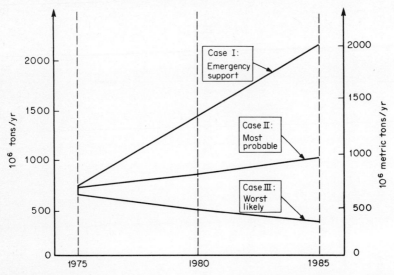

Fig. 7-15 Industry projections of U.S. annual production of coal under three policy alternatives. (U.S. Department of Commerce, National Technical Information Service, PR/237 605, p. 8. Prepared by Science Communications, Inc., October 1974.)

total energy input to the industrial sector is anticipated to increase from 24.1 quads (25.4 × 10^{15} kJ) in 1974 to approximately 43.3 quads (45.8 × 10^{15} kJ) in 2000. This represents an average annual growth rate of 2.3 percent.

Recent studies[10] indicate that significant changes in energy inputs are probably in future years. Referring to Table 7-7, notice that, while the petroleum share is expected to rise from a 1974 base year, coal is expected to decline by half that amount for the comparable period through 2000. The gaseous fuels share over this period is expected to fall by less than half its 1974 percentage, while the electrical share should roughly triple.

The industrial sector also has the potential to become less energy-intensive in the years ahead due to the following factors:

1. Increased efficiency and utilization potentials.
2. The changing mix of manufactured items away from energy-intensive programs.
3. The apparent shift to increased use of electricity, with a related shift of losses to electric utility and secondary energy (i.e., synthetic gas and liquid) sectors.

It is also interesting to consider industrial sector conversion losses due to the probable assignment of secondary energy sources and their attendant conversion losses, as shown in Table 7-8. Industrial sector consumption, as a percentage of the

TABLE 7-8 Energy Consumption Responsibility of the Industrial Sector, 1974–2000, 10^{12} Btu and 10^9 kJ

	1974	1980	1985	2000
Conventional primary fuels:				
Coal	4,208 (4,439)	4,800 5,064	4,930 5,202	5,910 6,235
Petroleum	6,044 6,376	7,500 7,912	8,370 8,830	10,370 10,940
Natural gas	11,129* 11,741*	10,000 10,550	9,500 10,022	9,000 9,495
TOTAL	21,415 22,592	22,300 23,526	22,800 24,054	25,280 26,670
Secondary energy sources:				
Synthetic liquids			130 137	1,230 1,298
Synthetic gas			240 253	2,260 2,384
Utility electricity	2,665 2,812	3,600 3,798	5,260 5,929	14,680 15,487
TOTAL	2,665 2,812	3,600 3,798	5,990 6,319	18,170 19,169
Conversion losses assigned from secondary energy sources:				
Synthetic liquids			55 58	530 559
Synthetic gas			110 116	970 1,023
Utility electricity	5,551 5,856	6,960 7,343	10,650 11,236	24,385 25,726
TOTAL	5,551 5,856	6,960 7,343	10,815 11,410	25,885 27,308
Total energy consumption responsibility of sector	29,631 31,260	32,860 34,667	39,605 41,783	69,335 73,147
Total U.S. gross energy consumption	73,121 77,142	87,040 91,827	103,540 109,234	163,430 172,419
Sector consumption as a percent of total	40.5	37.8	38.2	42.4

*Includes 34 × 10^{12} Btu (36 × 10^9 kJ) of hydropower not otherwise shown.
SOURCE: L. J. William, "Preliminary Forecast of Energy Consumption Through 1985," Electric Power Research Institute, Palo Alto, Calif., March 1976.

U.S. gross energy total, has minimal fluctuations. It drops slightly following 1974, then rises to a value of 1.9 percent above the comparable 1974 level in 2000.

7-13 LIMITATIONS IN FORECASTING

Energy planners are often forced to operate without reasonable assurance that plans made today as a result of national study and analysis of available pertinent data will be relevant for more than a few years.

Sophisticated forecasting techniques spread throughout industry during the fifties and sixties. They were built upon scientific methods, theories of business cycles, advanced mathematical techniques, game theory developed during the thirties and forties, and the well-known Delphi techniques developed at Rand Corporation. Recent modifications to the Delphi methodology, designed to minimize personal persuasion and maximize convergence of views, have recently been extended to social, political, and economic forecasting problems.

Other useful methods in the technological forecasting arsenal, in addition to trend extrapolation, include narrative methods (such as morphological analysis or decision trees) and cross-impact analysis.[11] Why, then, have forecasts become irrelevant? And why has distrust of their use reached almost epidemic proportions in recent times? In the fifties and sixties, the future was easier to understand; it was much like yesterday. In short, forecasting was based on extrapolation of relevant historical data.

Nowhere is the crystal ball more cloudy than in forecasting our probable energy needs. Change and uncertainty are more the rule than the exception. In searching for predictive techniques for the evaluation of rational alternatives, corporate planners must recognize that trying to identify the future in times of uncertainty can be misleading and, therefore, potentially disastrous. We are often surrounded by contradictory signs and ambiguous "omens." Only recently have we come to realize that our future needs may be better served by the use of scenarios.

7-14 USE OF SCENARIOS IN FORECASTING

Webster defines a scenario as "an outline or synopsis of a play." In their book *The Year 2000,* Herman Kahn and Anthony J. Wiener define a scenario as a "hypothetical sequence of events constructed for the purpose of focusing attention on causal processes and decision points." The use of multiple scenarios in forecasting is aimed at freeing the forecaster from the need to formulate a particular sequence of events or trends. Instead, the forecaster can focus on their probable ranges. This is made possible by developing a theme based on given sets of relevant details sufficient to formulate probable demographic changes, technological developments, political events, social trends, and economical variables.

Scenarios of the future take on a variety of lengths, formats, and topics. In recent years, scenarios portraying U.S. and world energy problems have tended to dominate the literature. The now famous "Project Independence Report,"[2] published by the Federal Energy Administration in 1974, consisted primarily of an analysis of three scenarios of future U.S. energy needs. This report provided three different estimates, accordingly, of projected 1985 U.S. energy consumption levels. The first scenario described the domestic energy supply-versus-demand situation based on the federal government taking no action. In that scenario, domestic oil imports remained high. The second scenario described the domestic energy supply-versus-demand situation based on government action limited to restraints in energy demand. The last scenario portrayed the U.S. domestic energy supply-versus-demand situation based on government actions to stimulate domestic energy supplies. In the second and third scenarios, a reduction in oil imports resulted. However, varying social and economic consequences contributed to differing results.[12]

Another well-publicized scenario dealing with U.S. energy policy was set forth in the final report of the Ford Foundation's Energy Policy Project.[13] This report also attempted to describe in detail the U.S. energy picture with three scenarios. One of these, the "historical growth" scenario, portrayed the consequences of continued U.S. energy growth at a 3 to 4 percent annual rate. A second "technical fix" scenario portrayed the consequences of employing energy saving technologies, which might result in the above historical growth rate being reduced to about 1.7 percent per year. Finally, a controversial third scenario portrayed the consequences of a future zero-energy growth rate. The prospect of an energy policy aimed at no growth in domestic energy demand after the mid-1980s has aroused considerable public speculation and criticism.

Unfortunately, all the above scenarios focused on various energy policy options and are not particularly useful for industry planning efforts. Recently, various industries have recognized these limitations. Many have used in-house expertise to develop their own corporate projections. Scenario development for corporate planning falls into one of two general categories:

1. Scenarios of the environment
2. Scenarios of the firm

The first category deals with defining and then limiting the future universe that corporations may encounter. In such scenarios, one addresses, at a very macrolevel of description, external factors, such as:

1. *Economic*: rate of inflation, labor cost, raw material availability, cost of capital
2. *Political*: allocation, regulation, relevant legislation

The second scenario category deals with responses both within and external to the firm. It addresses such issues as:

1. Cash flow
2. Capital expansion needs
3. Tax consequences
4. New investment opportunities
5. Evaluation of market response to new products
6. Perceived consumer needs and demands
7. Energy or fuel consumption

Among the major industrial companies that have reported the successful use of such mixed scenarios are General Electric, Shell Oil, and Monsanto.[13]

A key decision to be made in the development of meaningful energy scenarios for industry concerns the variety and range of issues to be addressed. The selection of viable scenario themes and the delineation of a time period to be covered each represent major challenges to the industrial planner. The literature describes many methods (such as matrices and dynamic, or econometric, models). Avoidance of too much detail is recommended. Multiple (alternative) scenarios of equal length and development, reflecting the required breadth of viewpoint, are necessary to keep the scenario team to a small, manageable size.

Energy impacts may be short- or long-range. The purpose of scenario development in energy planning is to identify and examine those corporate decisions that may be affected in the near future. Long-range objectives can always be realigned with updating. As a management tool scenarios also provide the means for possible management interaction in examining the many energy risks and options often overlooked in more conventional forecasting methods.

Let us consider a typical scenario development process.[14] Our example involves the substitution of domestic coal for current industrial operations utilizing either gas or oil. We will consider three general coal supply scenarios for the purposes of illustration:

Case 1: Expansion of coal production facilities under government-sponsored emergency conditions.

Case 2: Expansion of coal production facilities on a 10-year projected basis and under conditions most likely to occur.

Case 3: Expansion of coal facilities under the worst possible circumstances, which are recognized as only marginally possible.

Case 1

Let us begin by assuming that coal is priced at $20/ton ($23.3/t), that is, 12,000 Btu (12,670 kJ) bituminous coal at the mine head in constant dollars. Assume that virtually unlimited capital is available at a subsidized rate of 5 percent. Assume further that only present environmental constraints on mining are enforced, transportation is readily available, and no serious depression exists. Mine, Health and Safety Act regulations remain as written and enforced in 1974. Current technology is employed, and industry may burn 2 to 3 percent sulfur coal without scrubbers, either through relaxation of controls or through perfection of gas cleanup technology. Under these government-sponsored emergency circumstances, industry production in millions of tons per year is projected as follows:

	1975		1980		1985	
East	595	586	650	639	750	738
West	158	155	320	314	459	451
TOTAL	753	741	970	953	1,209	1,189

Capital requirement will be approximately $30/ton ($33/t) annually of new capacity. As a result, about $6 billion/ton ($6.6/t) of new capital is required each year. (Note: It should be recognized by industry planners, however, that under these circumstances case 1 goals may be unattainable.)

Case 2

It is assumed that adequate load capital will be available at a prime rate of 10 percent, and that high-grade 3 percent sulfur content sells for $11/ton ($12.3/t) bituminous coal. Transportation will present problems. Furthermore, requirements for reclaiming strip mining land will become increasingly stringent as local resistance in the West is sustained; OSHA enforcement also will become stricter. Old facilities will be subject to intermittent control standards and can burn both high- and low-sulfur coal, if desired. New units will be equipped with SO_2 recovery units and burn high-sulfur coal only. However, current technology will dominate. Under these circumstances, coal production in millions of tons per year is estimated to be:

	1975		1980		1985	
East	585	575	620	610	700	689
West	120	118	192	189	313	308
TOTAL	705	693	812	799	1,013	997

(Note: We assume that capital requirements would increase from about $1.3 billion to $3.8 billion. This capital requirement may not permit coal to increase its contribution to national fuel consumption appreciably.)

Case 3

Under this case, capital remains tight and the government offers no special concessions to energy-related projects. Furthermore, there is effective resistance to Western strip mining. The shortage of rail transport persists and intermittently worsens. Even with current technology, there appears to be no effective cleanup system for SO_x. To complicate matters, more safety regulations are implemented.

Environmental protection codes are enforced as written in 1974, and coal used will have to contain less than 0.75 percent sulfur. Mine head prices in the East would range from $1.60 to $2.00/$10^6$ Btu ($1.51 to $1.90/GJ), with an average of approximately $1.75/$10^6$ Btu ($1.66/GJ) or $42/ton ($46.3/t).

Under case 3 circumstances, coal operators will be forced out of production unless they have low-sulfur coal, limited to about 0.6 percent sulfur content. Coal production in millions of tons per year is projected as follows:

	1975		1977		1980		1982		1985	
East	460	453	330	325	275	270	200	197	185	182
West	80	78	100	98	115	113	140	138	165	162
TOTAL	540	532	530	423	390	383	340	335	350	344

Figure 7-15, which summarizes these three cases, can be used by corporate planners to assist in making decisions on future process fuel substitutions. It should be apparent, however, that the key assumptions regarding the handling of sulfur in coal are critical in comparing the three cases, because the sulfur problem has not been quickly solved through combustion gas cleanup technology. Liquefaction or gasification of coal may offer a timely solution, but the probable risks must be carefully assessed.

DESIGN CONSIDERATIONS FOR INDUSTRIAL SECTOR CONSERVATION

The designer of new industrial facilities must take into account the possibilities for intraindustry fuel substitution. Representative mathematical curves which portray energy substitution by fuel versus prices for the probable range of interest may be of help. Also useful are curves of energy substitutions by fuel versus time.

Designers have devised many specific design improvements for the pulp and paper, the iron and steel, the cement, the chemical, and the petroleum-refining industries.

Finally, the industrial designer must take into account carefully created scenarios regarding future energy availability and demand.

SUMMARY

A great many conservation opportunities exist in the industrial sector of the economy. Despite serious logistical and political problems which may slow the gradual shift from oil and gas back to coal, significant intraindustry fuel substitution possibilities remain. Promising new techniques, using mathematical curves and the value-added principle, are being developed for planning and evaluating such substitution possibilities.

Important strides toward conservation are being made in the pulp and paper, the iron and steel, the cement, the chemical, and the petroleum-refining industries.

Promising new techniques are expected to materially improve energy consumption patterns in evaporation and distillation processes. These areas of the chemical industry will benefit from new vapor recompression techniques, multiple reboiling systems, and, possibly, from new developments such as the Peltier-effect diffusion still (PEDS).

Forecasting future energy needs in industry is important, but there are many pitfalls in developing reliable predictive techniques. The use of carefully reasoned scenarios can be a great help.

NOTES

1. "Intra Industry Capability to Substitute Fuels," Federal Energy Administration Report No. FEA-E1-50034, October 1974.
2. "Project Independence Report," Federal Energy Administration, November 1974.
3. S. H. Nelson and R. S. Goldman, "Use of Fuel Cells in District Heating Systems," Energy and Environmental Systems Division, Argonne National Laboratory, Chicago, Sept. 29, 1977.
4. R. L. Shaner, "Energy Scarcity: A Process Design Incentive," Chem. Eng. Prog. (May 1978).
5. H. A. Huckins, "Conserving Energy in Chemical Plants," American Institute of Chemical Engineers, 1978.
6. William C. Petterson and Thomas A. Wells, "Energy-Saving Schemes in Distillation," Chem. Eng. (Sept. 26, 1977).
7. M. Meckler, "Use Peltier Heat Pumps to Improve Process Separation Availability," Proceedings of the 14th Intersociety Energy Conversion Engineering Conference, Boston, August 1979.
8. M. Meckler, "Peltier Effect Heat Pump Systems," Paper No. 73-WA/PID-3, ASME Winter Meeting, Detroit, Mich., November 1973.
9. "Upgrading Existing Evaporators to Reduce Energy Consumption," Energy Conservation, U.S. Department of Commerce No. COO/2870-2, 1977.
10. L. J. William, "Preliminary Forecast of Energy Consumption Through 1985," Electric Power Research Institute, Palo Alto, Calif., March 1976.
11. T. J. Gordon and O. Helmer, "Generation of Internally Consistent Scenarios Through Study of Cross-Impact," presented at the Industrial Management Center, New York, Jan. 27–30, 1969.

12. Walter G. Dupree, Jr., and S. Corsentino, "United States Energy Through the Year 2000," Bureau of Mines Report No. BuMines SP-8-75, December 1975.
13. S. D. Freeman (ed.), *A Time to Choose: Final Report by the Energy Policy Project of the Ford Foundation,* Ballinger, Cambridge, Massachusetts, 1974.
14. Rene D. Zenter, "Scenarios in Forecasting," *Chem. Eng. News* (Oct. 6, 1975).
15. Meckler, *op. cit.*
16. Ibid.

Environmental Impacts of Energy Technology

Reduction in terrestrial carbon reservoirs since 1850 has resulted in atmosphere carbon dioxide increases.

Minze Stuiver

8-1 ENVIRONMENTAL IMPACTS OF FOSSIL-FUELED ELECTRIC PLANTS

Recent studies[1] by the Electric Power Research Institute (EPRI) suggest that commercially available, large-scale, low- and medium-Btu gas processes for coal gasification, such as the Lurgi, Koppers-Totzek, and Winkler systems, will not be cost-competitive in the near term. During the early seventies, a number of processes aimed at extracting and employing clean fuels derived from coal were examined. Major developmental risks and associated costs were brought to the attention of interested governmental agencies. Much time was lost in attempting to fund the development costs for these synfuels plants. Meanwhile, costs escalated rapidly, no doubt aggravated in part by rising environmental concerns. A comparison of the capital costs of 10 different clean ways to combust coal, also developed by EPRI[1] is listed in Table 8-1. It features both conventional and combined-cycle steam plants.

Four such clean fuel option systems which hold promise for the 1985–1990 time frame (and probably not before) are:

1. Pressurized fluidized-bed combustion (PFBC)
2. Atmospheric fluidized-bed combustion (AFBC)
3. Low-Btu gasifier combined cycle (GCC)
4. Solvent-refined coal (SRC)

This premise is based on comparisons with stack gas cleanup (SGC) systems on a conventional power plant in terms of overall cost and environmental

TABLE 8-1 Capital Costs for Ten Clean Ways to Burn Coal

	Heat Rate		Base Cost, $/kW	Contingency, %	Uncertainty, %	Total Cost* $/kW
	Btu/kWh	kJ/kWh				
		Conventional Steam Plants				
Low-sulfur coal	9,000	9,495	290	+10	±10	375–460
High-sulfur coal with alkali scrubbing	9,500	10,022	PP: 290 SR: 50 340	+10 +20	±10 ±20	485–625
High-sulfur coal with regenerative scrubbing	10,000	10,550	PP: 290 SR: 150 440	+10 +20	±10 ±20	575–740
Atmospheric fluidized-bed combustion	9,500	10,022	PP: 340	+20	+25, −15	450–665
Solvent-refined coal	9,000 BC: 10,000	9,495 10,550	PP: 290	+15	±15	375–500
Petroleum-type fuel	9,000 BC: 13,400	9,495 14,137	PP: 190	+10	±10	250–300
Low-Btu gas, moving-bed, dry ash Lurgi process	BC: 13,600	14,348	PP: 190 SR: 390 580	+10 +20	±10 ±15	760–1000
Medium-Btu gas, slagging moving-bed process	BC: 11,300	11,921	PP: 190 SR: 255 445	+10 +20	±10 +25, −15	585–800

Process	BC/Other	PP/SR	±	Range
Low-Btu gas, atmospheric, two-stage entrained process	BC: 10,600	PP: 190 SR: 210 400	+10 +20	±10 +25, −15	525–710
Medium-Btu gas, pressurized, two-stage entrained process	BC: 9,800	11,183	PP: 190 SR: 155 345	+10 +20	±10 +25, −15	490–600

Combined-cycle Plants

Process	BC/Other	PP/SR	±	Range
Petroleum-type fuel	7,500 BC: 11,200	7,912 11,816	PP: 160	+15	±15	185–250
Low-Btu gas, moving-bed, dry ash Lurgi process	7,500 BC: 9,500	7,912 10,022	PP: 160 SR: 335 495	+15 +20	±15 ±15	650–875
Medium-Btu gas, slagging moving-bed process	7,500 BC: 9,100	7,912 9,600	PP: 160 SR: 215 375	+15 +20	±15 +25, −15	490–895
Low-Btu gas, atmospheric, two-stage entrained process	7,500 BC: 8,400	7,912 8,862	PP: 175 SR: 180 355	+15 +20	±15 +25, −15	460–650
Medium-Btu gas, pressurized, two-stage entrained process	7,500 BC: 8,150	7,912 8,598	PP: 160 SR: 130 290	+15 +20	±15 +25, −15	375–530

*Includes IDC and start-up at 30 percent (except 22 percent for combined-cycle petroleum-type fuel plant).
PP = power plant.
SR = sulfur removal system.
BC = basis coal (coal conversion and power generation).
SOURCE: R. C. Rittenhouse, "Clean Fuels from Coal: Finding the Right Combination," *Power Eng.* (October 1977).

intrusion. It is recognized that the available SGC systems are generally lower in efficiency than the above suggested technologies.

For example, EPRI compared a 1000-MW power plant with a stack gas scrubber (SGC) and one with a GCC system. In Table 8-2 notice that, in almost every category, the SGC system consumed either more fuel or generated more pollutants than the GCC alternative or both.

There are presently no federal standards that apply selectively to synthetic fluids. Furthermore, current practices in diffusion modeling, meteorological and air quality monitoring for class I increments, as defined under the Clean Air Amendments of 1976,[2] may be unenforceable. Therefore, we have to question the available technology or significant pollution deterioration requirements we are likely to face by 1985. Environmental groups are currently pressuring the Environmental Protection Agency (EPA) to narrow the so-called "90 percent rule" to mean that 90 percent of the sulfur would require removal regardless of the fuel used. Such a ruling, if applied literally, would lock in SGC systems and eliminate incentives for the use of clean fuel alternatives. Flexibility in meeting the intent of applicable clean air standards is essential if we are to explore various alternative clean fuel options to determine the best system.

Five years after Congress ordered the cleanup of U.S. air to protect public health, about two-thirds of the U.S.'s 247 air quality control regions had failed to meet one or more of the clean air standards. Nevertheless, EPA Administrator Russel E. Train said that, even though there is still a long way to go, the United States had made significant progress in cleaning up the air. The standards established are for total

TABLE 8-2 Environmental Impact of 1000-MW Power Plants

	Coal-fired Boiler + Stack Gas Scrubber	Gasification-combined-Cycle Power Plant
Coal consumption, lb/kWh (kJ/kWh)	0.92 (0.42)	0.78 (0.35)
Limestone required, lb/kWh (kJ/kWh)	0.12–0.15 (0.05–0.07)	0 (0)
Flue gas volume, scf/kWh (m^3/kWh)	125 (3.54)	125 (3.54)
NO$_x$ emissions, ppm	400–500	10–40*
SO$_2$ emissions, ppm	200 (20)†	10–200
Particulate emissions, gr/scf (g/m^2)	0.01–0.05 (0.02–0.11)	Negligible
Makeup water, gal/kWh (L/kWh)	0.6–0.65 (2.27–2.46)	0.4–0.45 (1.51–1.71)
Disposal land required, acres/1000 MW (ha)	1200–2400 (486–972)	200–500 (81–202)

*Texaco TPM date for oil gasification and low temperature tubine.

†20 ppm level probably attainable with current scrubber technology and low-sulfur coal. With high-sulfur coal, additional equipment may be needed at estimated cost of $60 to $90/kW.

SOURCE: R. C. Rittenhouse, "Clean Fuels from Coal: Finding the Right Combination," *Power Eng.* (October 1977).

suspended particulate matter (dust, smoke, and soot), sulfur dioxide, carbon monoxide, nitrogen dioxide, and oxidants.

The failure to reach standards has been generally attributed to the complexity of pollution abatement problems, confounded by the energy crisis and economic recession. In some cases, states don't have regulations that are tough enough. Often, court challenges have prevented the EPA and states from enforcing regulations.

Yet, there have been noticeable improvements. Sulfur dioxide concentrations nationally were reduced in 1975 about 25 percent over 1970. The national average for particulate matter dropped 14 percent between 1970–1971 and 1972–1973, and the trend continued at least into 1974. About 78 percent of the 20,000 major stationary sources (industries, power plants, municipal incinerators) are complying with regulations. Levels of carbon monoxide exceeding the eight-hour standard for auto emissions have declined more than 50 percent nationally. And concentrations of photochemical oxidants, particularly in Los Angeles and San Francisco, have shown improvement.

There is special concern about the particulate standards. Some of the new controls required to meet those standards fully might be very unrealistic in the near terms. A number of areas are expected to fail to meet the particulate standards even if there is full compliance with the emission limits for major stationary sources. This is because of the significant background level of particulates from the wear of vehicle tires, construction and demolition activities, and windblown dust from open lands and streets.

Sulfur dioxide emissions will be increased by the return of high-sulfur coal. If we are not careful, energy demands could unravel much of our progress to date. We must remember that clean air is not an aesthetic luxury; it is a necessity for public health.

Electric utilities are among the major contributors to air pollution. The EPA has estimated that the utilities were responsible for 50 percent of sulfur oxides, 20 percent of the nitrogen oxides, and 20 percent of particulate matter discharged nationwide in 1968. Most of this air pollution was the result of burning coal as a basic fuel. However, carbon monoxide from transportation sources amounted to over twice (by weight) the total pollution of the utilities.

Recent legislation specifically restricts the amounts of various air pollutants that electric utilities may emit per million Btu (1.055×10^6 kJ) of fuel consumed. Widely used by the electric utilities to remove particulate matter are cyclone separators and electrostatic precipitators. These devices can be used effectively in meeting the existing air quality standards for maximum acceptable particulate discharge rates. The presence of sulfur oxide in the boiler flue gases enhances efficient particulate removal. A method of removing sulfur from coal fuels involves the use of limestone or dolomite to absorb the sulfur oxides in the combustion chamber of the boiler. A wet-limestone scrubbing process is currently being developed for use in the combustion chamber of the exhaust stack to efficiently remove both the sulfur oxides and the particulate matter. Nitric oxides are the result of high-temperature, excess-air combustion. Lower furnace temperatures would reduce NO_x emissions, but they would also reduce plant efficiency, increase the emission of particulates, and tend to produce carbon monoxide emissions. Limited studies have indicated that the wet-limestone scrubbing process to absorb SO_2 may also reduce nitric oxide emissions. The removal of SO_x, NO_x, and particulate emissons from coal-burning power plants is interrelated. A well-designed, total emissions control system is required to reduce each of these components to acceptable discharge rates. The following case study illustrates the dimensions of the problem.

8-2 REPRESENTATIVE CASE STUDY

In a recent office project, we were faced with estimating the relative emission values for conventional oil- or gas-fired electric generation plants. These were plants without scrubbers but with supplementary individual oil-fired heating systems. The plants were comparable to oil- and gas-fired diesel engines or turbine-driven total energy plants capable of producing 100 MW of electricity and recoverable (waste) thermal energy. Using representative total energy scenarios taken from our office files, the following key parameters were assumed:

1. The conventional total energy plant case operates at an estimated 55 percent utilization.
2. Purchased electric utility power generation plants operate with an average heat rate of 11,380 Btu/kWh (12,006 kJ/kWh).
3. Indirect-type supplementary industrial heaters operate at an overall 60 percent thermal efficiency.

An additional scenario incorporating a solar total energy system of the type described in Chap. 5 was also constructed. It was based on an assumed Rankine-cycle, engine-driven induction motor-generator, operating with an overall cycle of 15 percent efficiency. It was also assumed to be operating within a system conservatively requiring 5 percent of the annual electrical output to be dedicated to parasitic system loads, such as pumps, miscellaneous motors.

Estimated emission factors[3] for each of the major pollutants, (organics, NO_x particulates, and CO) are shown in Table 8-3. They are converted for each pollutant to grams per kilowatthour of the fuel.

It was determined that the conventional total energy plant consumed approximately 73.5 percent of the fuel required by the electric utility at the point of generation. In Tables 8-3 through 8-6, note that the emission factors listed for internal combustion engines (applicable to diesel-driven total energy plants) appear rather high for each of the major pollutants when fired with diesel oil. Emission factors for

TABLE 8-3 Organics from 1 kWh of Fuel (grams)

Fuel	Resid. and Commer. Heat Low	High	Industrial Heat Low	High	Transport Low	High	Electricity Low	High
Gasoline					4.80	6.67		
Diesel					1.98	2.20		
Ref. make gas				0.004				
Aviation gas					6.13	6.55		
Jet fuel					4.20	4.33		
Distillate oil	0.027	0.045						
Residual oil			.027	0.048			0.0388	0.0516
Piped gas		0.0015	0.01	0.11	0.256	0.282	0.006	0.020
Liquefied petroleum gas					0.178	0.178		

SOURCE: E. J. List: "Energy Use in California: Implications for the Environment," EQL Report No. 3, December 1971, California Institute of Technology, Pasadena, Calif.

TABLE 8-4 Carbon Monoxide from 1 kWh of Fuel (grams)

Fuel	Resid. and Commer. Heat Low	High	Industrial Heat Low	High	Transport Low	High	Electricity Low	High
Gasoline					27.4	34.2		
Diesel					0.34	0.830		
Ref. make Gas			0.0042	0.0042				
Aviation gas					20.3	30.3		
Jet fuel					0.40	1.69		
Distillate oil	0.002	0.022						
Residual oil			0.002	0.020			0.0004	0.002
Piped gas	0	0.0006	0	0.0006	1.33	5.63	0	0.0006
Liquefied petroleum gas					2.56	2.56		

SOURCE: E. J. List: "Energy Use in California: Implications for the Environment," EQL Report No. 3, December 1971, California Institute of Technology, Pasadena, Calif.

TABLE 8-5 Oxides of Nitrogen from 1 kWh of Fuel (grams)

Fuel	Resid. and Commer. Heat		Industrial Heat		Transport		Electricity	
	Low	High	Low	High	Low	High	Low	High
Gasoline					1.40	2.41		
Diesel					1.34	2.48		
Ref. make gas				0.22				
Aviation gas					1.82	2.00		
Jet fuel					0.18	0.51		
Residual oil			0.74	0.83			0.37	1.23
Piped gas	0.164*	0.33	0.26	0.33	0.22	0.46	0.28	2.47†
Liquefied petroleum gas					0.20	0.43		

*A kitchen range actually has the lowest emission factor, 0.086 g/kWh.
†Exceptionally high figure from one particular power plant.
SOURCE: E. J. List: "Energy Use in California: Implications for the Environment," EQL Report No. 3, December 1971, California Institute of Technology, Pasadena, Calif.

TABLE 8-6 Particulars from 1 kWh of Fuel (grams)

Fuel	Resid. and Commer. Heat		Industrial Heat		Transport		Electricity	
	Low	High	Low	High	Low	High	Low	High
Gasoline					0.143	0.153		
Diesel					1.23	2.06		
Ref. make gas			0.152	0.152				
Aviation gas					0.150	0.150		
Jet fuel					0.148	0.795		
Distillate oil	0.089	0.221						
Residual oil			0.146	0.255			0.0126	0.216
Piped gas	0.0275	0.0293	0.0260	0.0278			0.0220	0.0232
Liquefied petroleum gas								

SOURCE: E. J. List: "Energy Use in California: Implications for the Environment," EQL Report No. 3, December 1971, California Institute of Technology, Pasadena, Calif.

gas- or oil-fired turbine-driven total energy plants were assumed to correspond to industrial heading values listed in Tables 8-3 through 8-6.

Employing averaged values for each of the respective pollutant high-low headings under Tables 8-3 through 8-6, estimated reductions (or increases) can be computed for each of the pollutants over those generated by the combined conventional oil-fired electric utility generation plants and supplementary industrial heating facilities. These were to be displaced by the equivalent performance of the comparable conventional or solar total energy plants.[4,5]

Results of a major study included the following:

1. Employing diesel engine–driven conventional total energy plants will not result in any savings in pollutant load over present operations. On the contrary, it tends to increase the pollutant load by a factor of approximately 4 : 1, despite the estimated 27 percent reduction in fuel input requirements.

2. Employing No. 2 residual oil in turbine-driven conventional total energy plans should result in an estimated savings in pollutant loading of approximately 4.7 tons/d (4.3 t/d).

3. Employing natural gas fuel in lieu of No. 2 oil to the turbine-driven conventional total energy plant should result in a pollutant load savings of approximately 14 tons/d (12.7 t/d).

4. Finally, if solar total energy plants were employed in lieu of conventional plants of the type described earlier, the estimated maximum savings in pollutants load would approach approximately 19.4 tons/d (17.6 t/d).

It was concluded that some mix of natural-gas- and residual-oil-fired turbine-driven and solar total energy plants could materially reduce the pollutant emissions for comparable utility generation features. Furthermore, the use of solar total energy plants apparently provides us with an additional 39 percent reduction in pollutant load. Therefore, solar total energy plants should be encouraged as the method having the least adverse impact on areas with limited available natural gas supplies.

8-3 NUCLEAR POLLUTION IMPACTS

What to do with the spent fuel elements from light water reactors is one of the most serious problems facing the nuclear industry. Currently, spent fuel is being stored mostly at reactor sites. If storage space is not expanded, available spent fuel storage space will be filled by 1980. The prospects for reprocessing remain dim, and the industry appears to be moving toward a de facto throwaway fuel cycle.

The annual discharge of spent fuel is expected to be 1722 t (1898 tons) of uranium by 1980, and 3594 t (3961.7 tons) by 1985. A maximum of 23,338 t (25725.7 tons) of uranium of spent fuel will be accumulated by 1985 if no reprocessing takes place. The amount of spent fuel will increase at the rate of 15 to 25 percent, annually.

No reprocessing plants are currently operating in the United States to handle the dramatic growth of spent fuel. Not long ago, the Federal Power Commission projected annual growth at 6.3 percent through 1985. The period from 1986 through 1995 is expected to experience a 5.7 percent growth rate per year. Others feel that the upward trend will be a bit less than 3 percent per year through the next 10 years.[1]

The Edison Electric Institute (EEI) reported electric output up to 6.3 percent for the 52 weeks ending December 25, 1976. This was echoed in individual reports from various utilities. The Federal Power Commission has also predicted regional power shortages by 1985.

The figures on new construction, based on reports from utilities, indicate nothing to brighten the prospects of heading off potential shortages. And, when the high costs of needed building facilities are considered, the picture does not grow brighter. The EEI estimates that $122 billion will be needed by investor-owned utilities for the next five years' commitments.[6]

Critics of load forecasts prepared by electric utilities suggest that such forecasts are, in a sense, "self-fulfilling prophecies" because of the monopoly power exerted by an individual utility over customers in its service area.

A review of forecasting methodologies used by 14 electric utilities in the Pacific Northwest revealed that relatively simplistic forecasting techniques had been employed, at least until recently when the large utilities started using more sophisticated econometric methods. The forecasting track record of these utilities up to the 1973–1975 period, when an economic downturn was accompanied by conservation measures, was reasonably good.[6]

8-4 ENVIRONMENTAL AND SITING LAWS

State environmental and siting laws now require comprehensive reviews of environmental impacts for all proposed power plants. Laws adopted to date fall into two general categories: (1) laws regulating the power plant siting of nuclear and fossil power plants and transmission lines and (2) laws requiring environmental impact statement (EIS) requirements for nongovernmental power plants.[7,8] States with legislation in the former category as of 1976 are indicated in Fig. 8-1. A tabulation of affected power and transmission facilities is indicated in Table 8-7.

States with legislation in the latter category, also as of 1976, are indicated in Fig. 8-2. Some confusion has arisen, largely due to uncoordinated efforts among various heads of government and, in some cases, duplicate federal review procedures. If state regulations expand into the area of power plant designs and operation, the implications for the future siting of power facilities could be far-reaching. For example, the design of so-called *power parks,* with installed generating capacities of from 10 to 50 GWe, has raised concern regarding possible microclimate weather modification

Environmental Impacts of Energy Technology   207

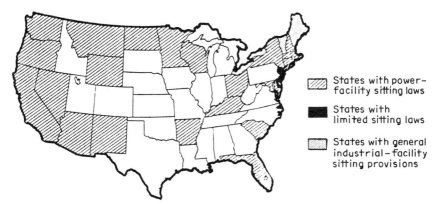

Fig. 8-1 States having siting regulations for power facilities. (Victoria A. Evans, "State Environmental and Siting Laws for Power Facilities," *Power Engineering,* August 1976.)

due to excessive localized heat and moisture loading. Presently, the maximum generating power plant site is approximately 3000 MWe. Studies[9,10] suggest that, at that level, heat dissipation does not cause major atmospheric effects.

Although the energy park concept has the advantage of siting away from populated areas, if we assume 33 percent overall plant efficiency for the generating units, from 20 to 100 GWe of waste energy must be discharged into the atmosphere. This

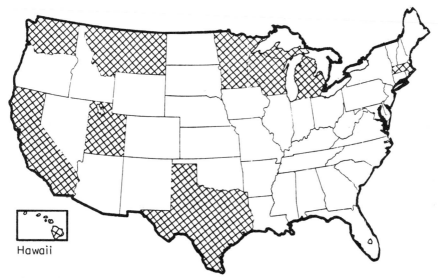

Fig. 8-2 States having environmental impact statement requirements for power facilities in 1976. (Victoria A. Evans, "State Environmental and Siting Laws for Power Facilities," *Power Engineering,* August 1976.)

TABLE 8-7 States with Siting Laws Regulating Power Facilities

	Facility and Minimum Size or Voltage Regulated	
State	Power Plant	Transmission Line
Arizona	100 MW	115 kV
Arkansas	50 MW	100 kV and 10 mi long, or 170 kV and 1 mi long
California	50 MW	Line from plant to interconnection within state
Connecticut	Any	69 kV
Delaware	Any*	Any*
Florida	Any	"Associated line"
Kentucky	Any	400 kV
Maine	1 MW	125 kV
Maryland	Any	69 kV
Massachusetts	100 MW	69 kV and 1 mi long
Minnesota	50 MW	200 kV and 1 mi long
Montana	50 MW	34.5 kV
Nevada	Any	60 kV
New Hampshire	50 MW	100 kV and 10 mi long
New Jersey	Any*	Any*
New Mexico	300 MW	230 kV
New York	50 MW	125 kV and 1 mi long, or 100 kV and 10 mi long
North Dakota	50 MW	200 kV and certain 69–200 kV
Ohio	50 MW	125 kV
Oregon	200 MW	200 kV
South Carolina	75 MW	125 kV
Vermont		
Washington	250 MW	200 kV
Wisconsin	300 MW	100 kV
Wyoming	100 MW	"Associated line"

*In coastal zone only
Permits are required for both nuclear and fossil plants in all 25 states.
SOURCE: Victoria A. Evans, "State Environmental and Siting Laws for Power Facilities," *Power Eng.* (August 1976).

would require a land area of from 5 to 100 km^2, assuming cooling towers are used. Attempts have been made at modeling[9] the dispersion of pollutants as well as the buoyancy and momentum of effluents in the plume.

In view of recent regulatory and environmental constraints, it is becoming increasingly difficult to locate plant sites near urban areas. Government regulations and controls are believed to have kept processors and refiners from committing large sums to plants that may take years to build. Instead, a number of smaller plants have been constructed.

Tougher environmental standards have prevented construction of the 15 major refineries listed in Table 8-8, and they played a major part in Dow Chemical's decision not to build in California.

California recently published a report[11] which points out that municipal solids, feedlot, agricultural, and timber wastes are not presently utilized, and that the environmental impact of soft technologies is significantly smaller than that of all hard technologies. Such biomass could be converted to yield synthetic or pyrolysis gas and thereby be available for the operation of fuel cells, which would partially replace combustion processes.

Urban sites will permit stimulation of the local economy if the pollution impacts can be overcome. Electrical generation methods which supply DC electric power directly would seem to couple best with the emerging fuel cell (direct chemical energy conversion) technology soon to be commercially available in a 5-MW size range.

8-5 ADVERSE WEATHER EFFECTS FROM THE COMBUSTION PROCESS[8]

The need to find alternative clean fuels is further accelerated by a recognition among many scientists that the oceans appear to be losing their ability to absorb increasing amounts of carbon dioxide being released through the combustion of fossil fuels. Worldwide destruction of forested areas, a principal source of stored carbon, through transfer of land into agriculture[12,13] is a related problem. Figure 8-3 illustrates variations in atmospheric carbon dioxide trends as measured at the Mauna Loa Observatory in Hawaii. The seasonal oscillations appear to be caused by removal of carbon dioxide through photosynthesis during the summer months and subsequent release during the fall and winter months.

Each year, the combustion of fossil fuels alone releases roughly 20 billion tons of carbon dioxide into the atmosphere. Wallace Broecker, a Columbia University geochemist, has estimated that for every tankful of gasoline purchased for an automobile, 300 to 600 lb of carbon dioxide is produced. It is estimated that the United States now generates 27 percent of the world's carbon dioxide. However, Dr. Rotty of the Institute for Energy Analysis[14] warns that our share over the next 50 years could drop to 7 percent of this total, as 58 percent of the total then will be produced by the now-developing countries of Asia alone. This adds another dimension to the U.S. energy policy problem; we need to set a meaningful example in energy conservation that can be emulated by other nations before it is too late.

It has been suggested that, with ever larger amounts of carbon dioxide remaining in the atmosphere, a global warming trend should be apparent by the end of this century. This phenomenon, popularly referred to as the *greenhouse effect*, results

TABLE 8-8 Refineries Planned But Not Built Because of Environmental Opposition

Company	Location	Size 10^3 L/d	Size bbl/d	Final Action Blocking Project
Shell Oil Co.	Delaware Bay, Del.	23,850	150,000	State legislature passed a bill forbidding refineries in coastal area.
Fuels Desulfurization	Riverhead, N.Y.	31,800	200,000	City Council opposed project and would not change zoning.
Maine Clean Fuels*	South Portland, Maine	31,800	200,000	City Council rejected proposal.
Maine Clean Fuels*	Searsport, Maine	31,800	200,000	Maine Environmental Protection Board rejected proposal.
Georgia Refining Co.*	Brunswick, Ga.	31,800	200,000	Blocked through actions of Office of State Environmental Director.
Northeast Petroleum	Tiverton, R.I.	10,335	65,000	City Council rejected proposal.
Supermarine, Inc.	Hoboken, N.J.	15,900	100,000	Withdrawn under pressure from environmental groups.
Commerce Oil	Jamestown Island, R.I.	7,950	50,000	Opposed by local organizations and contested in court.
Steuart Petroleum	Piney Point, Md.	15,900	100,000	Rejected by St. Mary's County voters by referendum on July 23, 1974.
Olympic Oil Refineries, Inc.	Durham, N.H.	63,600	400,000	Withdrawn after rejection by local referendum.
C. H. Sprague & Son	Newington, N.H.	7,950	50,000	Voted down in community vote on June 28, 1974.
Belcher Oil Co.	Manatee County, Fla.	31,800	200,000	Voted against in referendum of Sept. 10, 1974.
In-O-Ven	New London, Conn.	63,600	400,000	Abandoned because of opposition from state government and citizens' groups.

*Maine Clean Fuels and Georgia Refining Co. are subsidiaries of Fuels Desulfurization.

SOURCE: James H. Prescott, "Small is In, Big is Out in Oil-Refining World," *Chemical Eng.,* p. 84 (Oct. 10, 1977).

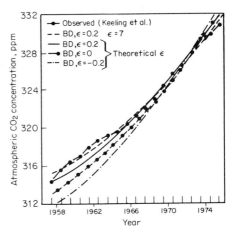

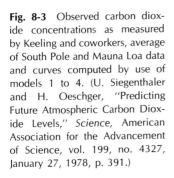

Fig. 8-3 Observed carbon dioxide concentrations as measured by Keeling and coworkers, average of South Pole and Mauna Loa data and curves computed by use of models 1 to 4. (U. Siegenthaler and H. Oeschger, "Predicting Future Atmospheric Carbon Dioxide Levels," *Science*, American Association for the Advancement of Science, vol. 199, no. 4327, January 27, 1978, p. 391.)

from the ability of carbon dioxide molecules to absorb radiated (infrared) heat and reemit it back toward the surface of the earth.

This is analogous to a glass enclosure through which sunlight passes and is then trapped. While there is much debate regarding the consequences of probable weather changes, there is agreement that a change in a few degrees (°F) could drastically alter conditions on earth.[14] The rate of change in the atmospheric concentration of carbon dioxide in the last 10 years is alarming. From the start of the Industrial Revolution (1850) through 1976, human activities increased the CO_2 concentration from approximately 290 ppm to 400 ppm, of which approximately 19 ppm has been attributed to the period 1957–1976. See Fig. 8-4 for an illustration of this. Assuming a continuation of this recent trend, by 2020 we can anticipate a doubling of current CO_2 concentration levels in the atmosphere. Should this occur,

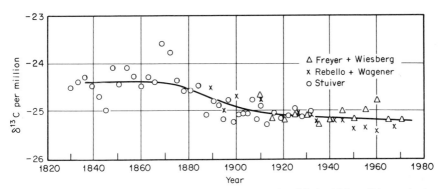

Fig. 8-4 Change in atmospheric $\delta^{13}C$ after correction for fossil fuel addition. ("Atmospheric Carbon Dioxide and Carbon Reservoir Changes," *Science*, American Association for the Advancement of Science, vol. 199, no. 4326, January 20, 1978, p. 255.)

we can expect a 2 to 3°C (3.6 to 5.4°F) average rise in worldwide climatic conditions. Dramatic changes on polar regions, rising ocean levels and rainfall patterns cannot be predicted with any certainty. This means that greater reliance on coal, a major source of CO_2, as proposed by the Carter administration, could have greater long-term, worldwide environmental impact. One thing is clear: We are pumping carbon dioxide into the atmosphere at a faster rate than our planet's regulatory mechanisms can cope with. This is illustrated in Fig. 8-5.

Nuclear energy has been targeted for a 50 percent level as a primary energy source by 2050. However, an innovative approach to the problems plaguing the nuclear industry seems overdue. It has been suggested by Cesare Merchetti,[12] of the International Institute for Applied Systems Analysis, that we decouple the energy conversion process from the sociosystem, and that while doing so we continue to pursue a continuation of the economy of scale in electric power generation. One suggestion is the use of high-temperature, gas-cooled nuclear reactors deployed on energy islands, or floating offshore platforms. Instead of delivering electric power by conventional transmission lines, these reactors could generate liquid hydrogen which could be transported by tanker to energy centers. Employing liquid air as an energy carrier would permit the use of ocean thermal gradient power plants, ideally through isothermal compression and subsequent power extraction by isothermal expansion to obtain useful work at the point of use.

Energy development and conservation are intimately related. We must begin now to accelerate the development of renewable, nonfossil sources for our intermediate

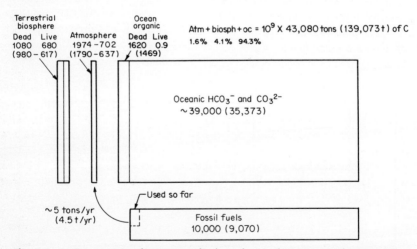

Fig. 8-5 Major reservoirs of importance for the carbon cycle. Units: 10^9 short tons (10^9 t, or metric tons) of carbon. ("Atmospheric Carbon Dioxide and Carbon Reservoir Changes," *Science,* American Association for the Advancement of Science, vol. 199, no. 4326, January 20, 1978, p. 254.)

and long-range energy needs, while encouraging controlled fossil fuel growth only as necessary to satisfy our near-term needs. Doubtless, this transition will cause some dislocation within our economy. It is likely to result in hardships for some Americans. The longer we delay, the weaker our dollar will be and the more difficult change will become. Failure to act soon could polarize our society and drain our ability to deal with the massive environmental consequences of unabated carbon dioxide levels in the atmosphere. Such an eventuality could make our current concern with pollution a trivial matter by comparison.

8-6 COPING WITH AIR POLLUTION REQUIREMENTS

With the EPA currently requiring that power plants switch to coal, it may be helpful to examine the design and economic impacts of compliance with governmental environmental requirements.

In past years, designers of coal-fired power plants started at the front end by selecting the boiler and pulverizer. They would then determine the feed capacity, based on the properties of the coal utilized. As a result of the Clean Air Act Amendments of 1977, designers now start with baseline site emission studies that establish the least limits of incremental additions to the surrounding air. Plans are based on predicted future regulatory constraints and the quality of the air monitoring equipment to be used. Refer to Table 8-9 to obtain some idea of the cost of compliance as reported recently by Indianapolis Power & Light. The effects of coal characteristics are dramatically illustrated in Table 8-10, which points up how three different coal variations, nominally referred to as A, B, and C, can affect air pollution equipment. Nuclear plants face increased scrutiny; they must provide improved methods to reduce the volume of concentrated solid and liquid radioactive wastes in chemical wastes, unexchanged resins, and sludge from filters. With respect to liquid discharge, the EPA is required under the 1977 Water Pollution Control Act Amendments to reexamine its 1984 list of 65 toxic substances and to issue effluent guidelines limiting such toxic substances in all major categories of industrial discharge.[15]

The designer of power plants is thus faced with a long list of choices in monitoring equipment. Positive control methods must be established for dealing with construction runoff, dust and noise control, cooling towers drift, and numerous other problems.[16-18]

Since Drs. Rowland and Molina first theorized that fluorocarbons might harm the ozone layer in 1974, the EPA has investigated fluorocarbon regulation. It was believed that the EPA would regulate refrigerants through the usual methods—allocation, ban, labeling, or containment. It now appears that the EPA is investigating economic regulation to increase the price of fluorocarbons as an alternative to

TABLE 8-9 Expenditures for Compliance with Government Environmental Requirements as Reported by Indianapolis Power & Light

Power plant	Air Quality Control				Water Quality Control			
	(1) Expended through 1977	(2) Expended in 1977	(3) Budgeted Expenditures 1978–1982	(4) Total (1) + (3)	(1) Expended through 1977	(2) Expended in 1977	(3) Budgeted Expenditures 1978–1982	(4) Total (1) + (3)
Petersburg	$64,209,000	$22,344,000	$ 53,221,000	$117,430,000	$36,701,000	$16,965,000	$25,605,000	$ 62,306,000
H. T. Pritchard	4,898,000	−27,000	96,000	4,994,000	1,677,000	−4,000	219,000	1,896,000
E. W. Stout	7,680,000	283,000	111,000	7,791,000	6,107,000	81,000	1,053,000	7,160,000
Perry K	6,995,000	−225,000	1,151,000	8,146,000	720,000	−2,000	55,000	775,000
Perry W	2,033,000	10,000		2,033,000				
Future location	912,000	760,000	55,302,000	56,214,000	3,217,000	674,000	25,258,000	28,475,000
Environmental consulting services	90,000	18,000		90,000				
Capacity cost	8,058,000	7,244,000	6,961,000	15,019,000	10,734,000	7,880,000	11,312,000	22,046,000
Total capital investment	$94,875,000	$30,407,000	$116,842,000	$211,717,000	$59,156,000	$25,594,000	$63,502,000	$122,658,000

Total capital investment, air and water quality control, through 1977: $154,031,000
Total capital investment, air and water quality control, through 1982: $334,375,000
Additional environmental aesthetic costs (substation landscaping, underground lines, steel transmission towers, etc.), 1969 through 1977: $ 7,101,000

*Additional generating capacity required to operate environmental control equipment.

SOURCE: R. C. Rittenhouse, "Coping with Pollution Control Requirements for Power Plants," *Power Eng.*, p. 43 (July 1978).

TABLE 8-10 Estimated Quantity of Flue Gas Desulfurization Wastes and Ash from Selected Coals as Used in a 1000-MW Power Plant

	A	B	C
Ash content of coal, %	15	8	15
Sulfur content of coal, %	3.5	0.8	2.0
Heating value of coal, Btu/lb (kJ/kg)	12,500 (29,125)	8,500 (19,805)	12,500 (29,125)
Annual coal use, 10^6 tons/yr (10^6 t/yr)	2.63 (2.38)	3.85 (3.49)	2.63 (2.38)
Sulfur emission stands	1.2 lb/10^6 Btu (0.52 kg/10^6 kJ)	50% removal	1.2 lb/10^6 Btu (0.52 kg/10^6 kJ)
Annual sludge production, ton/yr dry (t/yr)	315,000 (285,705)	35,000 (31,745)	160,000 (145,120)
Annual ash production, ton/yr dry (t/yr)	395,000 (358,268)	310,000 (281,170)	395,000 (358,265)
Annual total solid wastes, ton/yr dry (t/yr)	710,000 (643,970)	345,000 (312,915)	555,000 (503,385)

SOURCE: R.C. Rittenhouse, "Coping with Pollution Control Requirements for Power Plants," *Power Eng.*, p. 44 (July 1978).

conventional banning or containment of the product.[19] It is thought that economic regulations would artificially raise refrigerant prices enough so that use of refrigerants would decrease.[19-22]

8-7 CROSS-MEDIA ENVIRONMENTAL IMPACTS

In the past, pollution regulations have often been written and enforced without a sufficient analysis of the secondary environmental impacts likely to result. A measure designed to improve air quality will prove counterproductive, for instance, if it results in increased contamination of the land or water. Such cross-media environmental impacts must be taken into account if we are to have a sound environmental policy. Trade-offs must be weighed in the regulation of specific pollutants, and secondary effects must be carefully and systematically categorized.

Often, the technologies most efficient in reducing harmful emissions are also the most energy-intensive. The increased energy use in such a system may have substantial cross-media impacts which may negate the positive value of the emission control process. The best interests of the overall environment are not being served through the stringent control of a single pollutant if energy impacts and secondary environmental impacts are not being taken into account.

DESIGN CONSIDERATIONS AND THE ENVIRONMENT

The industrial designer must be aware of the growing maze of regulations pertaining to the environment. In planning for power plants using coal, advanced scrubbing techniques should be explored. Where possible, solar generation should be investigated. Problems of siting must be dealt with. The designer of power plants must also cope with environmental problems such as construction runoff, dust and noise control, and cooling towers drift. Cross-media environmental impacts must also be considered.

SUMMARY

Increasingly, problems of energy utilization are becoming intertwined with aspects of the environmental situation. Much legislation, both on the national and the state level, has been enacted regarding the construction and operation of electric utilities. Nuclear power is particularly restricted by governmental regulation, and problems of radioactive waste disposal remain to be solved.

The siting of new power plants are affected by laws regulating the actual siting and by laws requiring environmental impact statements. Such laws are making it increasingly more difficult to locate plant sites near urban areas.

Cross-media environmental impacts must be taken into account if we are to create a sound environmental policy.

NOTES

1. R. C. Rittenhouse, "Coping with Pollution Control Requirements for Power Plants," *Power Eng.* (July, 1978).
2. "Costs of Environmental Regulation Draw Criticism, Formal Assessment," *Science*, **201** (July 14, 1978).
3. E. J. List, "Energy Use in California: Implications for the Environment," EQL Report No. 3, December 1971, California Institute of Technology, Pasadena, Calif.
4. W. S. Duff and W. W. Shaner, "Solar Thermal Electric Power Systems: Manufacturing Cost Estimation and Systems Optimization," American Society of Mechanical Engineers, 76-WA/HT-14.
5. M. Meckler, "Promising Solar Energy Applications in Large Building HVAC Systems," *ASHRAE J.* (November 1977).
6. F. C. Olds, "Regulatory Growth: Impact on Power Plant Planning and Construction," *Power Eng.* (May 1977).
7. N. R. Dibelius and W. R. Marx, "Status of Environmental Legislation Affecting Fossil Fuel Fired Power Generating Equipment," ASME Paper No. 72-FT-23.
8. Victoria E. Evans, "State Environmental and Siting Laws for Power Facilities," *Power Eng.* (August 1976).
9. Chandrakant M. Bhumralkar and John A. Alich, Jr., "Meteorological Effects of Waste Heat Rejection from Power Parks," *Power Eng.* (August 1976).
10. David G. Jopling, "Large Capacity Plant Sites: Problems and Opportunities," *Power Eng.* (January 1976).
11. "Distributed Energy Systems in California's Future: Interim Report," vol. 1, U. S. Department of Energy, Report No. HCP/P7405-01, March 1978.
12. G. M. Woodwell, R. A. Whittaker, W. A. Reiners, G. E. Likens, C. C. Delwiche, and D. B. Botkin, "The Biota and the World Carbon Budget," *Science*, **199** (Jan. 13, 1978).
13. Dewey M. McLean, "A Terminal Mesozoic 'Greenhouse': Lessons from the Past," *Science*, **201** (Aug. 4, 1978).
14. U. Siegenthaler and H. Oeschger, "Predicting Future Atmospheric Carbon Dioxide Levels," *Science*, **199** (4327) (Jan. 27, 1978).
15. B. T. Lewis and J. P. Marron, "Special Purpose Rooms and Their Environment," *Facilities and Plant Engineering Handbook*, McGraw-Hill, New York, 1973, sec. 5, chap. 5.
16. John Papamarcos, "Stack Gas Cleanup," *Power Eng.* (June 1977).
17. William E. Archer, "Flyash Conditioning Update," *Power Eng.* (June 1977).
18. Allan M. Teplitzky, "Estimating Cooling Tower Sound Emission Levels," *Power Eng.* (September 1977).

19. "Dr. Edwards' Air Cycle Heat Pump, ROVAC, May Be the Answer to Fluorocarbon Pollution," *Build. Sys. Des.* (February/March 1977).

20. Robert P. Mader, "Economic Control: The Next Stage in Fluorocarbon Regulations?" *Air Cond. Refrig. Bus.* (February 1978).

21. "Industry Voice Must Be Heard on Fluorocarbon Issue," *Air Cond. Refrig. Bus.* (November 1977).

22. Robert P. Mader, "The Great Fluorocarbon Debate," *EPRI J.*, **2**(2) (March 1977).

U.S. Energy Policy at the Crossroads

We cannot afford to fail in reducing our oil imports. What is at stake is the survival of free democratic systems.

James Schlesinger

The tendency to import oil-related products is not expected to reverse until perhaps the mid-1980s. We cannot ignore the implications of this, for the United States and the industrialized nations as a whole. Figure 9-1 clearly shows how five major Western nations compared in their ability to cut costly oil imports from 1970 to 1976. Particularly alarming is the apparent inability or unwillingness of the United States to reduce its dependency on foreign oil while the other major industrialized nations are clearly doing just that.

In an attempt to allay growing public concern, administration experts have claimed that the United States can afford to run a trade deficit so long as foreign investors (and their governments) are willing to offset the outflow of U.S. dollars with equal amounts of capital inflow from overseas.

We are reminded that in the first 100 years of its existence, the United States imported more goods and services from abroad during 69 of these years. The rest of the time, exports exceeded imports only incrementally. Yet, the United States flourished and expanded. It did so because foreigners were investing in U.S. railroads, cattle ranches, and a myriad of industrial enterprises which provided a foundation for subsequent U.S. industrial expansion. During the 25-year period following 1850 alone, the cumulative deficit exceeded $1.6 billion, while long-term foreign investment grew to approximately $1.4 billion by 1869, thus generally offsetting the trade deficit for that period.

The situation reversed following 1900. During the early twentieth century and until recent years, the dollar value of U.S. exports almost always exceeded imports.

After World War II, large U.S. investments in the form of direct government aid, the Marshall Plan, and as private capital spending by business helped

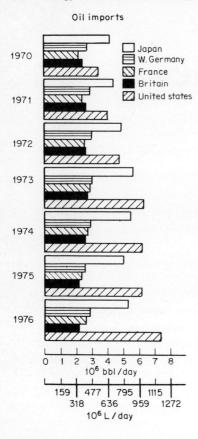

Fig. 9-1 Black gold: how five major Western nations have fared in controlling costly oil imports. (*Los Angeles Times*, June 22, 1978.)

restore the war-ravaged industrial base of Europe and Japan. Similar U.S. investment contributed also to the development of raw-material export industries in Latin America and Asia.

In 1970, total U.S. investments overseas amounted to just under $170 billion as compared to $97.7 billion of foreign investments in the United States.

Given the complexity and sophistication of today's world commerce, we can hardly draw comfort from such comparisons. The outward flow of U.S. dollars in investment payments alone, caused primarily by the high cost of imported oil and oil products, has contributed to the U.S. dollar's continuing weakness. If reliance on foreign oil continues unabated, U.S. inflation will result in a spiraling process, and further erosion of the U.S. dollar is inevitable. Lest investments pouring in from abroad be taken as an encouraging note, consider the following realities:

1. Jobs are not expanding in proportion to the investments.

2. The cost of social, educational, and medical services continues to erode real personal income.

3. The average age of the U.S. citizen continues to increase, placing a heavier burden on the younger working population.

4. The rising expectations of citizens in underdeveloped and competing industrialized nations can be expected to drive up the demand for finite world oil resources.

5. The technological gap, which has long created a demand for U.S. goods and services abroad, has been allowed to close as multinational corporate interests expand. This has resulted in a weak competitive position for many U.S. products.

6. The profits from foreign investments may not necessarily be recycled in the United States.

7. There is considerable doubt that major oil-producing countries, such as Saudi Arabia, will continue to expand their output beyond their ability to absorb or profitably invest their revenues.

8. Some experts predict a decline in the surplus normally generated by U.S. agricultural industries.

A recent rather sobering study by the Morgan Guaranty Trust Company of New York concluded: "To a considerable extent, the deficit is the product of U.S. inaction on energy policy. Probably the most constructive contribution the U.S. can make towards reducing its deficit is to develop and implement an effective energy program which tackles both the supply and demand sides of the problem, with the aim of limiting net oil imports."

The report went on to estimate that "taking account of offsetting exports of coal and some petroleum products, the bank's economists project the net deficit of nonnuclear fuels in 1977 as $43.5 billion, $47 billion in 1978, and $51 billion in 1979."

As pointed out by Morgan economist Rimmer de Vries: "We are going to have to have a very competitive economy [in world markets] to pay for this."

But can we have a competitive economy if the dollar cost of oil from OPEC nations continues to climb as a result of the deterioration of the dollar? This is a vicious circle. Our deteriorating dollar is now assisting the relative competitive positions of such nations as West Germany, Japan, and Switzerland. These nations and others with strong currencies have strengthened their positions, at least in terms of U.S. dollars, since their oil costs in dollar priced petroleum have decreased.

While some may justify the contemporary U.S. life-style, provided control inflows match outflows, the ability of the United States to promote its own interests must be questioned because of a combined dependence on overseas oil and capital. In this shrinking world, the threat of widespread disruption by foreign powers may be more psychologically damaging and compelling than the administration is presently willing to admit.

9-1 FUEL RESOURCES AND RESERVES

Resource terminology has not been standardized, and usage differs among the various energy industries and government agencies. When applied to minerals, the terms *resources* and *reserves* refer to two different concepts. The term *resource* represents a physical concept and denotes the quantity of a material that can be found within a given region. The term *reserves* refers to that portion of the resource which is known to a fairly high degree of accuracy to exist (as a result of field measurements) and to be profitably recoverable with current technology and under current economic conditions. Thus *resource* is a much broader term; it includes reserves as well as known deposits that cannot be profitably recovered, poorly mapped deposits, and undiscovered deposits that are hypothesized to exist on the basis of extrapolations from known fields.

Figure 9-2 provides a diagrammatic representation of the revised resources classification terminology employed by the U.S. Geological Survey (USGS) for petroleum. It depicts the relationship between resources and reserves. A similar approach has been used by the U.S. Bureau of Mines. The *measured reserves* category of the USGS is identical to the *proved reserves* category employed by the American Petroleum Institute and the American Gas Association. This term represents recoverable resources known to exist to a high degree of confidence. The term *remaining recoverable resources* refers to estimates of the fraction of the total hypothesized resource base that can be profitably extracted with current technology under current economic conditions. In a comprehensive study recently published by the Institute of Gas Technology (IGT), U.S. fossil fuel reserves were reported at 5.64 quads (5.95×10^{15} kJ), as compared to a worldwide figure ranging from 20.9 to 21 quads (22.04 to 22.15×10^{15} kJ). Additionally, the remaining recoverable fossil fuel in the United States is estimated to range between 28.6 and 45.6 quads (30.17 and 48.11 $\times 10^{15}$ kJ) as compared to world totals ranging from 107.5 to 122.5 quads (113.4 to 129.2×10^{15} kJ). If we assume the demand for fossil fuels will grow at a rate

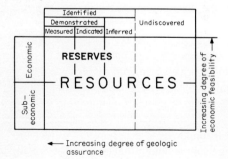

Fig. 9-2 Resource classification scheme employed by the U.S. Geological Survey. ("Coal Sources of the United States," January 1, 1974, U.S. Geological Survey Bulletin no. 1412.)

of 3 percent per year, based on a 10-year forward reserve allowance, our process reserves would last for 34 years and the remaining recoverable resources would last 88 years. If the annual growth rate were reduced to 2 percent per year, the corresponding values of proven reserves and remaining recoverable resources would increase, respectively, to 41 and 116 years.

IGT also reported that coal accounted for 84 percent of U.S. proved and currently recoverable fossil fuel resources. As a practical matter, U.S. coal reserves amount to 4.75 quads (5.01×10^{15} kJ). They considerably exceed the Middle East oil reserves, estimated to range from 1.86 to 2.08 quads (1.96 to 2.19×10^{15} kJ). The United States also has extensive amounts of syncrude from tar sands and oil shale which have not yet been commercially developed because of economic, environmental, and water supply considerations. Oil shale reserves have been estimated at 74 billion bbl (11.76 trillion L), as compared to the 1977 annual 3.0 billion bbl (0.47 trillion L) U.S. crude oil production total.

It must be recognized that the use of uranium will be strongly influenced by its manner of use and the relative cost of those ores which are recoverable economically. IGT estimated U.S. proved reserves of U_2O_8 of the $15/lb ($33/kJ) to be approximately 0.172 quad (0.181×10^{15} kJ), if employed in burner reactors without the benefit of plutonium recycle. However, if the ore were employed in breeder reactors, the latter value would soar to 12.9 quads (13.61×10^{15} kJ). Proven uranium reserves in the up–to–$30/lb ($66/kJ) category, however, can be figured as roughly double those in the $15/lb ($33/kJ) class.

Yet, in spite of the vast U.S. fuel resources available, the demand for energy and the mix of primary energy resources is not only highly dependent upon world petroleum costs but also on the rate of economic activity and population growth. The Stanford Research Institute (SRI) Energy Center[1] cited in 1972 the then anticipated energy consumption for 1990: 116.3 million bbl/d (18.5 trillion L/d) of Free World demand for oil, and 214 million bbl/d (34. trillion L/d) of oil equivalent for total Free World energy demand.

SRI now estimates the latter values, respectively, at 68.2 million bbl/d (10.8 trillion L/d) (or a 41 percent reduction) and 156.7 million bbl/d (24.9 trillion L/d) (or a 27 percent reduction) of oil equivalent. The reduction is a result of the estimated impact of energy conservation measures and the substitution of other energy resources. If such efforts can be intensified in the near term, it is possible that adequate oil supplies may be available through 2000. The demand for coal, natural gas, and uranium is nevertheless likely to be accelerated. While much attention has been focused on coal as a fuel, coal will also be used increasingly as a petrochemical source. Presently, coal-derived chemicals total 2.4 billion lb/yr (1.1 billion kg/yr), principally as a by-product in the manufacture of metallurgical coke. Yet the use of coal for petrochemicals and as a fuel will be affected by environmental restrictions, mining costs, and certain raw and resource material costs, such as imported oil.

The picture is further clouded by the fact that the manufacture of petrochemicals from oil or natural gas appears more expensive than from coal. Coal's composition and production costs are key factors in its suitability for petrochemicals. Yet the use of coal for chemical feedstock, particularly for ammonia in agricultural areas adjacent to producing coalfields, can be expected to grow in the years ahead. The coal currently recoverable, as well as the estimated total remaining recoverable values of coal assigned to energy production, must reflect this market condition.

9-2 PROGNOSIS ON NEW LONG-RANGE ENERGY RESOURCES

When we look about, we find plenty of oil around. Petroleum storage tanks in October 1977 held 970 million bbl (0.54 trillion L) of crude oil and fuel, or a 52-day supply, an inventory roughly 11 percent higher than a year earlier. Considering the rising flow of oil from Alaska's North Slope, we have to conclude that, for the near term, there is plenty of oil. The problem is one of policy. We must develop a program for placing the United States on a more self-sustaining course in the years ahead. Do we have the will, as a nation, to cut down our enormous oil imports? In 1977, these stood at 46 percent of total U.S. total needs, up an alarming 11 percent from 1973. The U.S. level was in sharp contrast to most other industrialized nations. It suggests a less adequate conservation effort by U.S. consumers than by those in Europe and Japan. See Fig. 9-1 for details. But what of our domestic oil and gas production potentials? Alaska's North Slope alone is expected to deliver an additional 500,000 bbl/d (79.5 × 10^9 L/d) in 1978, bringing the Trans-Alaskan oil line through-put to 1–2 million bbl/d. An equivalent displacement of foreign oil and dollar outflow is realized provided:

1. Domestic production levels remain the same.
2. Domestic demand is essentially static.

Yet domestic production levels from the lower 48 states continue to decline. Unless something is done to further stimulate U.S. oil and gas production, we can expect world oil output within the next few decades to follow the U.S. pattern. By 1990 or sooner it may begin to peak or perhaps decline.

Thus, a worldwide scarcity of petroleum threatens to arrive before we have developed alternate energy sources of equivalent capacity. Administration critics claim that over 20 years of low prices have left 98 percent of the prospective segments untouched by drilling. With respect to natural gas, they have identified four general categories of potential drilling sites which could be exploited if prices could be raised: frontier areas, deep geological basin areas, special category areas, and low-production areas.

Frontier areas are at remote distances from existing pipelines. Since no immediate market is available, long-term investment for development has not been justified.

Deep geological basin areas are locations where the largest reserves of natural gas are expected to come from depths below 15,000 ft (4572 m). In a recent year, only 404 such deep wells were drilled (1 percent of the total of 39,875 wells) because of the uncertainty of gas prices sufficient to provide an investment return. Many of these areas already produce oil and gas from shallow wells. Production from deeper zones is common in such areas when the higher drilling costs can be met with adequate product prices.

Special category areas include such regions as the Texas Gulf Coast, where 105,000 trillion ft^3 (2971 trillion m^3) of gas has been calculated to exist in high-pressure, hot salt water at 8000 to 25,000 ft (2438 to 7620 m). If U.S. gas consumption more than doubled to 50 trillion ft^3/yr (1.41 trillion m^3/yr) 10 percent of this special gas would supply the needs of the entire nation for 200 years. Its development awaits the assurance of proper pricing. These prices should still be below the cost of imported liquefied gas. The delivery system is in place. The nation's vast interstate pipeline system has cost over $50 billion to build, an expense which is paid by each consumer in proportion to his fuel use. The pipeline already crosses most of the areas where gas can be produced. Furthermore, full use of the complex pipeline network can be accomplished nationally just as it has already been done where flexibly priced intrastate markets have encouraged tremendous natural gas production increases.

Finally, low-production areas are locations where wells may produce slowly for many years or require expensive special recovery treatment. Such wells have not been economical under past price regulations. The Fort Worth Basin of Texas is such an area; unregulated prices there provided the incentive to increase drilling activity by 600 percent in four years, swelling gas supply 3000 percent in parts of the area. Other areas, such as Ohio with its undeveloped gas from shales, may show similar increases with the proper pricing of natural gas.

A recent study by the National Petroleum Council indicated that oil and gas production has been established in only about 50,000 mi^2 (129,605 km^2) (less than 2 percent) of the 3,000,000 mi^2 (7,776,300 km^2) identified as having oil and gas potential. A large part of established production is from shallow zones, with the deeper zones for natural gas remaining virtually untested. Tremendous quantities of new natural gas will be produced from the 98 percent of untested potential sediments as natural gas prices rise to the market value of other fuels.

The referenced study divided the United States into 11 regions for analysis of oil and gas potentials and available delivery systems currently in place. Distribution of promising sediments for U.S. oil and gas is illustrated in Fig. 9-3.

The estimated cubic miles of potentially favorable sediments in each of the 11 areas are listed below:

	Prospective Basins		Promising Sediments	
	Square Miles	Square Kilometers	Cubic Miles	Square Kilometers
1. Alaska	368,000	953,893	934,000	3,897,844
2. Pacific Coast states	126,133	326,949	251,508	1,062,133
3. Western Rocky Mountains	222,750	577,390	175,150	730,950
4. Northern Rocky Mountains	368,000	953,893	595,556	2,485,423
5. W. Texas/E. New Mexico	289,760	751,087	283,800	1,184,377
6. Western Gulf Basin	862,603	2,235,953	453,705	1,893,626
7. Midcontinent states	278,600	722,159	290,200	1,211,086
8. Michigan Basin	122,000	316,236	37,000	154,411
9. Eastern interior states	166,154	430,688	203,774	850,406
10. Appalachian states	130,000	336,973	305,000	1,272,850
11. Eastern Gulf/Atlantic Coast	268,000	694,683	558,700	2,331,612
TOTAL:	3,202,000	8,299,904	4,088,437	10,062,196

The Carter administration has taken bold leadership in calling for measures that will reduce our reliance on foreign oil. Only time will tell how effectively and maturely we, as a nation, rise to this challenge. The policy choices between energy conservation and resource development will continue for some time, and some vacillation may result. We are encouraged by improving prospects for isolation of our long-range energy problem and discouraged by the long-term aspects for solutions to growing long-range environmental problems. The key years lie immediately ahead. We must respond intelligently to the near-term energy problem. Continued efforts made through energy conservation only give us the time to apply the great American genius to the challenges of long-term resources and alternate energy supplies. At an annual meeting of the International Energy Agency (IEA) in Paris, the IEA, in its closing communique, said: "Unless present energy policies are strengthened there is a serious risk that, as early as the 1980s, the world will not have sufficient oil and other forms of energy available." This would result in severe economic, social and political consequences throughout the world. Nine new energy research and development agreements were signed, covering programs that will ultimately cost more than $130 million and bring to a total of 28 the number of such projects launched by the IEA since it began in 1974. The new agreements pertain to projects in the fields of coal research, solar power, fusion power, geothermal energy, wind power, and hydrogen production from water. Ultimately, it will take a worldwide effort to meet the challenge of an energy shortage and obtain some formula for setting a ceiling on oil imports among the IEA nations. IEA nations have urged that the United States reduce its oil imports 15 percent by 1985. Unless this is done, a significant world shortfall is anticipated. Its effects could be calamitous. The effects of policy on this goal are clearly apparent from a review of Table 9-1, which illustrates differing views on the projected U.S. energy supply in 1985. The table is based on the work of H. T. Franssen, of the U.S. Congressional Research Service, as

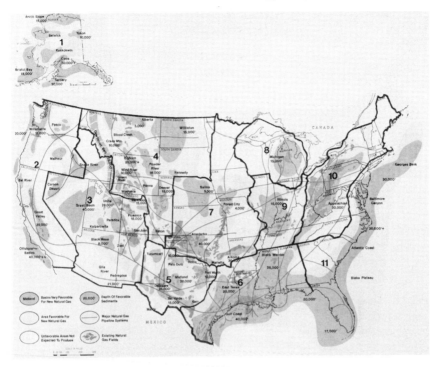

Fig. 9-3 Analysis of U.S. oil and gas potentials and available delivery systems. (Pitts Energy Group, Dallas, Texas.)

reported in November 1977 by the House Commerce Committee. This table compares Franssen's predictions with those contained in several earlier government reports and with the actual 1976 situation. Franssen concluded that several major oil finds may be necessary to achieve a 1985 production level of 11 million barrels per day. This suggests that our nation's long-term interests must not be made subservient to narrow party interests or regional viewpoints.

What may be at stake, ultimately, is our economic survival. In the first three months of 1978, the Commerce Department reported that the U.S. balance of payments, an important indicator of the country's economic position in the world, was in deficit by a record $6.95 billion.

The current accounts deficit was slightly larger than the $6.93 billion deficit in the fourth quarter of 1977, but much larger than the average of $2.7 billion in each of the three previous quarters. It was the seventh straight quarter of red ink.

The balance of payments is broader than the trade balance, which is reported every month. The balance of payments measures the money exchanged in trade, tourism, and service transactions with other countries. It also includes government payments abroad.

TABLE 9-1 Results of Summary of Estimated Import and Consumption of Oil, Gas, Nuclear, and Other Fuels in 10^6 bbl/d (10^{12} L/d) of Oil or Its Energy Equivalent

Year in Question	Whose Report	Assumptions	U.S. Oil	Imported Oil	U.S. Gas	Imported Gas	Coal Consumed	Nuclear	Other	Total
1985	Franssen of CRS (11/77)	Prices decontrolled for oil and new gas	9.5 (1.51)	13.5 (2.15)	8.1 (1.29)	1.0 (1.6)	8.9 (1.42)	3.0 (0.48)	1.9 (0.30)	46.1 (7.33)
1985	Carter administration (4/77)	Pre-Carter regulations prevail	11.3 (1.80)	11.5 (1.83)	8.2 (1.30)	1.2 (0.19)	10.9 (1.73)	3.7 (0.59)	1.7 (0.27)	48.5 (7.71)
1985	Carter administration (4/77)	Carter program is implemented	11.2 (1.78)	7.0 (1.11)	8.8 (1.40)	0.6 (0.09)	13.3 (2.11)	3.8 (0.60)	1.7 (0.27)	46.4 (7.38)
1985	General Accounting Office (7/77, 10/77)	Carter program is implemented	9.1 (1.45)	12.5 (1.99)	7.8 (1.24)	0.6 (0.09)	11.0 (1.75)	3.2 (0.51)		
1985	Federal Energy Administration (2/76)	Gradual deregulation of oil, gas prices	13.8 (2.19)	5.8 (0.92)	10.7 (1.70)	0.6 (0.09)	9.7 (1.54)	4.1 (0.65)	1.9 (0.30)	46.6 (7.41)
1985	Franssen of CRS (12/75)	Gas prices partly decontrolled	12.0 (1.91)	9.9 (1.57)	8.9 (1.41)	0.5 (0.08)	9.7 (1.54)	5.3 (0.84)	2.8 (0.45)	49.1 (7.81)
1976	Carter administration (4/77)	Actual	10.1 (1.61)	7.3 (1.16)	9.5 (1.51)	0.5 (0.08)	6.8 (1.08)	1.0 (0.16)	1.5 (0.24)	37.0 (5.88)

SOURCE: *Energy User News,* **2**(46):4 (Nov. 21, 1977).

U.S. Energy Policy at the Crossroads 229

The current accounts trade deficit increased sharply from $10.2 billion to $11.2 billion in the first quarter, reflecting a larger increase in imports than in exports.

9-3 SULFUR DIOXIDE EMISSIONS

A key policy issue currently being reviewed concerns sulfur dioxide (SO_2) emissions from power plants and factories. Major economic considerations hang in the balance; as emission controls grow more stringent, states that still have relatively clean air will gain increasing flexibility in trying to foster economic development. Under strict controls, utilities and manufacturing companies wishing to build new plants would have to arrange trade-offs which would ensure that, after their plants came on-line, overall SO_2 pollution levels would actually decrease. In other words, those wishing to build new plants in areas with high pollution levels would be forced to pay for scrubbing equipment in existing plants. Since scrubbing systems, particularly for full scrubbing, are extremely expensive, companies would more likely build new plants in areas with low SO_2 pollution levels.

The EPA must carefully sift through such factors before it makes a definitive ruling on permissible sulfur dioxide levels. In deciding such issues as full scrubbing versus partial scrubbing, the EPA will be measurably affecting the economic future of the United States.

9-4 THE NATIONAL ENERGY ACT OF 1978

The Department of Energy (DOE) hopes that the National Energy Act (NEA) will significantly reduce our imported oil needs by 1985. Other primary aims of the act are to increase the use of fuels other than oil and gas, and to provide for more efficient and equitable energy use within the United States.

The NEA comprises five separate bills:

1. The National Energy Conservation Act of 1978
2. The Power Plant and Industrial Fuel Use Act of 1978
3. The Public Utilities Regulatory Policy Act
4. The Natural Gas Policy Act of 1978
5. The Energy Tax Act of 1978

In addition to numerous provisions dealing with residential units, the NEA will provide $900 million over the next three years to assist schools, hospitals, and public buildings in paying for energy audits.

Energy audits are required in all existing federal buildings. Retrofitting is mandated so that, by 1990, all federal buildings will be operating at maximum efficiency. All new federal buildings must be designed with peak energy efficiency in mind.

Up to $100 million will be spent in the next three years to demonstrate solar technology by installing solar systems in federal buildings. This program is designed to stimulate the solar equipment industry, help lower its costs, and make solar systems more attractive commercially. For much the same reasons, the federal government is authorized to spend $98 million over three years to purchase photovoltaic energy devices for use in federal buildings.

All companies which consume more than 1×10^{12} Btu/yr (1.05×10^{12} kJ/yr) in one of the 10 most energy-consuming industries must report their energy consumption figures to the DOE annually. Such companies must also describe what actions they are taking to conserve energy.

NEA directs the DOE to establish target levels for increased industrial utilization of recovered materials. The DOE is also directed to evaluate pumps and motors to determine if new performance standards are necessary.

The Energy Tax Act provides a variety of tax credits for investment in energy conservation. An additional 10 percent credit is provided for investment in:

1. Boilers and other combusters which use coal or an alternative fuel
2. Equipment used to produce alternative fuel
3. Pollution control equipment
4. Equipment for handling or storing alternative fuels

These credits, not available to utilities, are designed to provide a major economic boost for the coal conversion regulatory program.

A refundable tax credit is available for investment in equipment which uses solar or wind energy to produce electricity, to heat and cool, or to provide hot water. This credit, also, is not available to utilities.

Other tax credits cover:

1. Equipment, such as heat exchangers and recuperators, designed to improve the heat efficiency of existing industrial processes.
2. Equipment used for the extraction or production of shale oil and geopressurized methane.
3. Equipment used for recycling waste materials.

New industrial oil or natural gas boilers, on the other hand, will not be entitled to an investment tax credit. They will be subject to straight-line depreciation unless federal or state air pollution regulations preclude the use of coal. If existing oil- or gas-fired boilers are retired earlier than originally anticipated, their useful life may be decreased. They are then subject to straight-line depreciation over that shortened remaining life.

The Energy Tax Act establishes a 10 percent depletion allowance for geopressured natural gas wells drilled between October 1978 and December 1973. A 22 percent depletion allowance is provided for geothermal energy from 1978 to 1980. This allowance is lowered gradually to 15 percent for 1984 and beyond.

Blends of lubricating oil containing oil which has been rerefined become exempt from the federal excise tax of 6 cents/gal.

Under the Power Plant and Industrial Fuel Act, new electric utility generation facilities or industrial boilers with a fuel heat input rate of 100×10^6 Btu/h (105.5×10^6 kJ/h) or greater may not use oil or natural gas unless the DOE grants permission. Furthermore, the DOE is granted the authority to require facilities capable of burning coal to use coal when possible. The DOE may also require units not capable of burning pure coal to burn coal-oil mixtures.

Utility power plants are limited to using no more natural gas than they used during 1974–1976. They may not switch from oil to gas. And, barring certain exceptions, all natural gas uses in such facilities must cease by 1990.

To aid utilities in raising necessary funds for pollution control, an $800 million loan program has been established. Certain other programs of limited scope are included to aid the conversion to coal.

Voluntary standards are established under the Public Utility Regulatory Policies Act of 1978 for state regulatory authorities and nonregulated utilities. Rate design, time-of-day practices, seasonal rates, and other utility practices are dealt with under this legislation.

Rules favoring industrial cogeneration are also provided. Among other things, utilities will be required to buy or sell power to and from qualified cogenerators at fair and reasonable rates. The Federal Energy Regulatory Commission is empowered to make such rules.

A loan program to assist in the development of small hydroelectric projects has also been established.

9-5 FEDERAL REGULATORY POLICY

Mushrooming government regulations are now threatening to stifle competition in the energy field. As the regulatory web grows, smaller companies, which lack the legal expertise to traverse the maze of rules and regulations, often find themselves unable to compete with larger corporations, with their extensive legal departments and staffs of experts.

Obviously, some government regulation is necessary. A balance must be achieved, however, between free competition and the impulse to regulate. Too often, government regulations have had the effect of encouraging monopolistic practices rather than limiting them.

Regulatory policy must come to view environmental problems holistically. Secondary impacts and cross-media effects abound in the area of environmental pollution, and these effects must be taken into account by those who devise the rules. The trade-off decisions involved may be difficult to make, but we must at least recognize that such trade-offs exist. Ongoing research, to quantify and characterize these trade-offs in energy transportation, combustion, and in the emissions which ultimately result must be accelerated. By definition, such research will involve economics, value judgments of various kinds, and other nonscientific criteria.

9-6 THE RESEARCH AND DEVELOPMENT CRISIS

A major area where excessive regulation is having an adverse effect is the research and development (R&D) field. Inconsistencies in government regulations administered by different agencies have seriously hampered the chemical and drug industries in recent years, for instance. Such inconsistencies cause companies to abandon projects prematurely in favor of sticking with already accepted products.

Many factors are causing the current decline in R&D. Tough government regulation, especially on the part of OSHA and the EPA, is most often cited as a stumbling block. Money which might otherwise be spent on innovative R&D must be earmarked for nonproductive, or defensive, R&D.

Skyrocketing inflation, which has dramatically increased the cost of doing business, is a second reason given for the slowdown in R&D. The cost of building new plants has risen dramatically. The lower profit margins which have resulted help keep R&D spending in line.

A third factor is rising personnel costs. Average salaries for research scientists and engineers have risen rapidly during the last decade. Between 1971 and 1976, average salaries rose a whopping 53 percent.

Government policies have also contributed to the decline. Slumping federal spending for R&D, the instability of the dollar, and the lack of economic incentives such as tax credits, low-cost loans, and the accelerated depreciation of research installations have not helped.

Unlike other industrialized countries, the United States currently provides no direct assistance for nonmission-oriented technological development. It is here, in the area of basic research, where we are falling behind.

Tables 9-2 and 9-3 illustrate the problem graphically. Table 9-2 breaks down R&D spending by area of performance. Table 9-3 shows us what this spending is in constant dollars. The annual growth rate is also exhibited. In terms of constant dollars, R&D spending in the United States has actually fallen during the last 10 years at an average annual rate of about 0.2 percent.

A significant point of dispute among the experts pertains to the effect of antitrust enforcement on R&D. Some authorities feel that strict antitrust policies inhibit R&D growth, since most U.S. companies cannot band together in joint research ventures

TABLE 9-2 U.S. R&D Spending by Area of Performance (millions of $)

	1960	1967	1972	1977 (Est.)
Total R&D	$13,523	$23,146	$28,257	$40,800
Federal	$1,726	3,396	4,482	6,500
%	13	15	16	16
Industry	10,509	16,385	19,383	27,750
%	78	71	69	68
Universities	1,006	2,594	3,440	5,133
%	7	11	12	13
Other	282	771	952	1,417
%	2	3	3	3

SOURCE: National Science Foundation.

on major projects. This, it is felt, puts U.S. corporations at a disadvantage, since other nations, particularly Japan, allow and encourage such joint ventures.

Other experts believe that tough antitrust enforcement helps R&D growth because it encourages smaller companies to enter the field. These authorities point out that, according to a recent report by the Office of Management and Budget, nearly half the major innovations over a 20-year period were achieved by companies employing fewer than 1000 workers. Yet, the federal government awarded only 8 percent of its R&D monies to such firms.

A related problem concerns the nature of current R&D. Intense social and regu-

TABLE 9-3 U.S. R&D Spending in Constant 1972 Dollars (millions of $)

	1953	1961	1967	1972	1977(E)
Total R&D	$8072	$20664	$29291	$28257	$28879
Federal	4675	13351	18217	15795	15429
Industry	3813	6866	10303	11502	12393
Universities	122	238	437	575	625
Other	92	209	334	385	432

	Annual Growth Rate, %		
	1953–1961	1961–1967	1967–1977
Total R&D	11.4	6.0	(0.2)
Federal	14.0	5.3	(1.7)
Industry	7.6	7.0	1.9
Universities	8.7	10.7	3.6
Other	10.8	8.1	2.6

SOURCE: National Science Foundation.

latory pressures for rapid change are forcing companies to spend their R&D funds on defensive technology, which often results in the patching up of existing technologies rather than the development of new and superior technologies. This climate of pressure and stringent regulation tends to discourage competition, especially from the smaller, newer companies. The many complicated steps required for the approval of new products discourages new producers and, thus, indirectly enhances the value of established and accepted products.

Fortunately, the Carter administration and the Congress have become aware of the R&D crisis. The administration is reviewing its policy and is expected to recommend some Presidential-level options on ways to stimulate innovation. Congress's Office of Technology Assessment is also examining the problem. It appears likely that Congress will reduce capital gains taxes significantly in the near future. This is expected to increase the availability of venture capital for small companies developing new technologies. Congress may also use selective procurement procedures and provide assistance to small companies in meeting regulatory requirements. Another positive step that Congress could take would be to strengthen patent protection for individual inventors and small companies.

A bold approach which has been proposed is the establishment of local technical centers throughout the country. Such centers would provide technical information and a greater diffusion of existing technologies to small firms.

Another proposal which has been advanced would require corporations to license technologies to their competitors once the leading firms reach a certain market share.

Most important of all is the need to establish national R&D priorities. Promising areas of basic research must be identified. Our most critical needs should be defined. Then we can determine which institutions can best conduct research in specific fields. In short, what is called for is a comprehensive national science policy.

9-7 THE PRODUCTIVITY LAG

A problem related to the R&D crisis is the lack of growth in productivity. According to some measurements, American productivity growth has been falling steadily since 1966. Spending for productive, nondefensive R&D programs is known to increase productivity, so renewed efforts to spur R&D growth may help correct this deficiency.

Once again, excessive government regulations are one of the primary causes of the problem. A second factor is the extent to which industries can convert to electronic production processes. Electronic technologies are clean and energy-efficient. For the most part, those industries which have managed to use electronic processes extensively have also kept increasing their productivity.

High energy costs have also contributed significantly to the productivity problem. When energy was cheap, industries sought to replace labor with machinery, often with little regard for the machinery's energy efficiency. Now rising energy costs have forced some older plants to close altogether. In other plants, extensive patch-up

U.S. Energy Policy at the Crossroads   235

efforts are underway to increase energy efficiency. In many cases, new electronic technologies are the key components in such efforts.

9-8 SPECULATIONS ON THE ROAD AHEAD

In this book, we have attempted to examine energy technologies and techniques available currently or expected in the near future. Systems which require more research or which are not expected to be available for many years have been labeled as such. This is because, from a practical viewpoint, there is a great deal of difference between a technology expected to be on-line in five years and a technology which might come into being, say, by the end of the century.

The simple truth is that many of the more grandiose energy proposals and potential solutions being offered up today are impractical, either from an economic viewpoint or from the standpoint of engineering design. Those who promulgate such ideas without pointing out the obvious economic or engineering problems involved often help create a climate of unnecessary skepticism. This, in turn, may adversely affect our chances of achieving rational solutions to our energy problems.

Ultimately, the solutions will prove to be simple rather than grandiose. In the distant future we will learn to use hydrogen from the seas as a primary source of energy. A given weight of hydrogen can provide three times as much energy as the same weight of gasoline. When hydrogen is burned to produce energy, water is the only by-product left over. There are no pollutants or harmful emissions. It is even theoretically possible to make a synthetic gasoline from hydrogen and carbon dioxide. Such a gasoline would be nonpolluting; only carbon dioxide and water would be released during combustion.

Of course, many serious problems remain to be solved before such an energy source can be realized. Hydrogen gas is extremely light and takes up much room. Moreover, it is highly explosive.

Nevertheless, scientists, researchers, and engineers with imagination and ingenuity will ultimately prevail. Some day, perhaps, the energy problems we face now will belong to the distant past.

SUMMARY

Questions of energy policy directly affect us all. The vast quantities of oil we are importing place us at a strategic and economic disadvantage. Fuel resources and reserves are dwindling rapidly.

The National Energy Act of 1978 includes provisions dealing with residential units, mandatory energy audits for federal buildings, and large grants to help develop solar technology. Large energy-consuming industries are required to report their energy consumption to the DOE annually. A variety of tax credits are provided for investment in special machinery, solar or wind energy used to produce electricity, and other systems geared toward energy conservation.

Mushrooming government regulation and a lack of effective new research and development programs threaten to stifle advances in the field. Major efforts must be made to reverse these trends.

Ultimately, solutions to our energy problems will be found. Human imagination and ingenuity will prevail.

NOTES

1. George B. Dantizig and S. C. Parikh, "On a Pilot Linear Programming Model for Assessing Physical Impact on the Economy of a Changing Energy Picture," Systems Optimization Laboratory, Department of Operations Research, Stanford University, Palo Alto, Calif., August 1975.

Energy Conservation Checklist

Appendix A

GENERAL RECOMMENDATIONS

1. Perform an energy audit.
2. Make a periodic review of the energy used.
3. Hold regular meetings with building personnel on energy conservation matters.
4. Ask utility companies to explain current rate structure. If possible, get them to revise to a lower rate.
5. Examine solar energy possibilities regarding your building.
6. Examine computer use. Make sure the equipment and the operating personnel are energy-efficient in all respects.
7. Make sure office machines are off when not in use.
8. Turn off unneeded elevators after normal business hours.
9. Replace terminal dispatching with multiple zoning controls on elevators.
10. Install additional insulation to ceilings, roofs, and walls where needed.
11. Use light-colored roofing materials to reduce solar gain on buildings with air conditioning.
12. Apply solar film to windows to cut cooling loads.
13. Replace single glazing with double glazing where appropriate.
14. Fit windows with solar screens to reduce cooling loads.
15. Make sure caulking on windows and door frames is airtight.
16. Make sure curtain walls are properly sealed.
17. Reduce crackage between double-entry doors.
18. Apply weather stripping around doors and windows.
19. Fix broken windows.
20. Make sure warehouse doors and garage doors are closed as much as possible.

238 Energy Conservation in Buildings and Industrial Plants

21. Repair louvers, dampers, valves, piping, and insulation to conserve energy.

22. Make sure that the supply of building services matches the needs of the occupants.

23. Throttle back pumping systems when possible to reduce motor horsepower.

24. Install small supplemental systems for operation during off-peak hours.

LIGHTING AND POWER RECOMMENDATIONS

1. In public areas, decrease light levels to a point that provides basic safety.

2. Lower lighting levels to 50 footcandles (15.24 metercandles) at office work stations.

3. Lower lighting levels to 30 footcandles (9.144 metercandles) in general working areas.

4. Lower lighting levels to 10 footcandles (3.048 metercandles) in halls, corridors, and other seldom occupied areas.

5. Replace high-density desk lights with fluorescent lights whenever possible.

6. Use reflective covers or backers on fluorescent lights to make maximum use of light refraction.

7. Compare different types of light sources to find the most efficient type for any given application.

8. Measure correct lighting energy with an ESI footcandle calculator.

9. Use one 100-W bulb to replace two 60-W lamps. This will save 12 percent of previous usage while providing the same amount of light.

10. Whenever possible, use skylights for natural lighting.

11. Make sure lights are out in unused offices.

12. Curtail the use of neon signs.

13. When possible, have custodial personnel work during daylight to reduce the use of lights.

14. Monitor public lighting and curtail every other light when no one is in a given area.

15. Make sure all unneeded lights are turned off on weekends and during other nonworking hours.

16. Cover dark walls in offices and halls with light colors to increase reflected light.

17. Rearrange light switch system to allow for more flexible selective switching.

18. Use fewer lamps in light fixtures.

19. Use smaller-capacity ballasts.

20. Fit office windows with aluminum blinds to reflect light and heat back into occupied office space.

21. When air conditioning is in use, reduce lighting levels. About ½ W of air conditioning power is used for each 1 W of lighting.

22. Adopt a lighting maintenance program to ensure maximum efficiency for lighting systems.

23. In perimeter office spaces, use natural lighting whenever possible.

24. Cut back on decorative lighting.

25. In outdoor lighting, use photocells or timers.

26. Use motors which are properly sized.

27. Whenever possible, apply power factor correction.

28. Install equipment for demand lighting systems.

29. Install advanced lighting systems which maintain or enhance lighting levels while reducing electric consumption.

RECOMMENDATIONS FOR CONTROLS

1. Make sure all controls have been newly recalibrated.

2. To prevent tampering by unauthorized personnel, lock all thermostats.

3. Make sure thermostats are not located on cold walls, in drafty areas, or in areas with direct lighting radiation.

4. Whenever feasible, install individual room control.

5. In radiators using hand valves, install self-contained temperature control valves.

6. For optimum use of outdoor air for building cooling, install enthalpy control.

7. When practical, install a building automation system.

8. Use selective synchronization of dampers to achieve optimum fan discharge temperatures.

RECOMMENDATIONS FOR HEATING AND AIR CONDITIONING

1. Curtail unnecessary overtime operations of heating and air conditioning systems.

2. Use less heat in winter.

3. Lower air conditioning levels in summer.

4. Keep units and registers unobstructed.

5. At night, close drapes to help retain heat in offices.

6. In the morning, open drapes so that sunlight can help heat and light offices.

7. In summer, close blinds and drapes to keep coolness in.

8. From October to April, maintain thermostat controls at 68°F (20°C) in all public areas.

9. From May to September, maintain cooling controls at not less than 78°F (25°C).

10. Curtail heat and air conditioning in unoccupied spaces and storerooms.

11. Make sure fans, filters, coils, and other equipment are kept constantly clean.

12. Eliminate drafts wherever possible.

13. To reduce the stack effect, make sure all openings on the top floor are secure.

14. For good heat transfer, blow down domestic hot water.

15. To lower heat gains from direct sunlight, use heat-reflecting and heat-absorbing glass.

16. Use systems and equipment at full capacity whenever possible.

17. For more comfort in offices and lower heating fuel costs, increase moisture in the air.

18. In hot water systems, use additives that help dissipate heat.

19. Check efficiency labeling when buying new equipment.

20. Make sure heating system motors and pumps are clean.

21. Check into the possibility of using a lower wattage motor.

22. Check into possible use of a catalytic burner for more efficient heating energy use.

23. Consider a master control which regulates heating and cooling system demand according to external temperatures.

24. Use variable controlled heating systems in place of on/off burners.

25. To lower heat loss, make sure foundations are waterproof.

26. Eliminate humidity controls whenever possible.

27. Where applicable, take advantage of low-per-unit costs at high demand by using one master meter.

28. Since consumption for the present month is used to calculate demand charge units for the following month, ask tenants to refrain from turning on air conditioning systems until after the month's meter reading.

29. Reduce outside-air intake as much as possible during heating and cooling seasons.

30. Check into the possibilities of energy recovery units which transfer thermal energy to reduce fuel consumption and equipment capacity.

31. Avoid clogged air conditioning condensate drain pans by using an anticlog additive.

32. Install an automatic controller on fans for use when only a part of total cooling capacity is required.

33. Schedule cleanup periods so that lights and systems can be turned off earlier.

34. Choose the most efficient system start-up time.
35. Adjust constant volume, variable volume, or mixing box and multizone unit dampers to reduce leakage.
36. Whenever possible, do not use preheat coils.
37. Keep perimeter and interior systems from clashing with each other.
38. Make sure multiple chiller operation is working at peak levels.
39. Use cooling and heating system auxiliaries only when absolutely required.
40. When feasible, use return-air fans for heating only during unoccupied hours.
41. To lower fan horsepower, employ auxiliary air risers.
42. Make sure exhaust systems are only exhausting the required amount of air.
43. Lower exhaust-air quantities from toilets and laboratories.
44. Make sure toilet room exhaust fans only operate when occupants are present.
45. Lower the domestic hot water supply temperature.
46. To preheat domestic hot water, make use of water from the condenser.
47. To reheat air conditioning, also use condenser water.
48. Install solar collectors to heat domestic and process water where feasible.
49. Make sure hot water circulating systems are off at night and during weekends.
50. Use faucet sprayheads or smaller nozzles on water fixtures to raise the pressure while reducing the amount of water used.
51. Install infrared heaters in place of forced-air heaters where suitable.
52. Install direct-fired equipment in place of indirect-fired makeup air units.
53. Make sure ductwork and piping in unconditioned spaces are properly insulated.
54. Make sure boilers, furnaces, pipes, etc., are properly insulated.
55. Lower chilled and hot water flows.
56. Make sure pump impellers match load.
57. Replace three-way valves with variable-speed pumping and two-way operation.
58. Make sure expansion tank is of the proper size. Excessive water loss can result from undersized tanks.
59. If possible, shut down the boiler plant for the summer. Use small boilers and water heaters in its place.
60. Return condensate to the boiler.
61. Investigate the use of steam turbines for fan and pump drives wherever high-pressure steam is plentiful.
62. To eliminate fouling in boilers, chillers, and heat exchangers, make sure the water treatment system is operating properly.
63. Inspect cooling tower bleed-off regularly to make sure water and chemicals are not being wasted.

64. Make sure evaporative coolers, cooling towers, and air-cooled condensing equipment is operating at peak efficiency.

65. Make sure cooking equipment is operating at peak efficiency.

66. Perform periodic checks on steam traps to eliminate steam waste.

67. Use enthalpy control to conserve on cooling costs by bringing in outside air when possible.

RECOMMENDATIONS FOR COMBUSTION EQUIPMENT

1. Make sure possible negative pressure in building is not lowering combustion efficiency.

2. Make sure no blockages or unwanted draft conditions are present in chimneys and flues.

3. Make sure all combustion surfaces are clean.

4. Make sure fuel-air ratios are adjusted properly.

5. Install power burners in place of atmospheric burners.

6. On industrial furnaces, install pressure controls.

7. Make sure equipment is not being overfired.

8. Make sure furnace linings are replaced regularly.

9. Keep production equipment preheat times to a minimum level.

10. Keep equipment temperature levels low when production stops for long periods.

11. Make sure curing and drying ovens are shut down when not in use.

12. Make sure there are no cracks in ovens and furnaces.

13. Use waste heat to preheat combustion air.

RECOMMENDATIONS FOR INDUSTRIAL PLANTS

1. Make sure gaskets and seals are secure around refrigerated spaces.

2. Make sure there is proper insulation of furnaces and all piping containing hot or cool materials.

3. Make sure there are no leaks in process steam pipes and hoses, as well as in air compressor systems.

4. Make sure motors, conveyors, etc., have adequate lubrication.

5. Check temperature control equipment to make sure it is properly gauged.
6. To improve heat transfer, all buildup on boiler surfaces should be removed.
7. Use waste heat to preheat boiler feedwater and materials which require heat.
8. When feasible, reduce unnecessary ventilation.
9. Use waste materials as supplemental fuels when possible.
10. Secure recommendations from qualified engineers regarding which ventilating, heating, and air conditioning equipment is best suited for a given application.
11. Investigate the placement of infrared heaters in high-bay areas and warehouses.
12. Make regular checks on stack temperature and possible gas concentrations from combustion processes.
13. Make use of automatic combustion system controls and time clock controls where applicable.
14. Stress boiler efficiency and conduct frequent boiler maintenance.
15. Check furnaces for hot spots which may signal lining deterioration.
16. To preheat incoming combustion air, use recuperators and combustion equipment where applicable.
17. Divert some electrical use to off-peak hours when voltage remains high.
18. Return steam condensate to boilers.
19. Optimize the plant power factor.
20. To identify leaks in building surfaces and equipment, use an energy leak thermometer.
21. To discover unneeded heat loss in fire-tube boilers, use a polarized scale which indicates steam pressure and corresponding temperature.
22. Keeping covers closed on vats and tanks lowers evaporation losses.
23. Fifty percent of the air or more can be saved by using push-pull ventilation on open-surface tanks.
24. When feasible install immersion heaters.
25. When feasible use cold water detergent in washing processes.
26. Use fans with extended ducts to warm floor areas. This cuts down on air stratification in the plant.
27. When workers are far apart, use spot heating or cooling.
28. When possible, use evaporative cooling for personnel.
29. Avoid the use of compressed air for cooling personnel.
30. When possible, replace compressed air usage with pressure blowers.
31. Keep compressed air at lowest suitable pressure.
32. Used oil can be burned in the boilers or reused after refining.
33. Make sure trucks and forklifts are being used efficiently.

34. Make sure heat and smoke relief vents are kept closed during winter.

35. Investigate the use of wastewater for spraying down roof to reduce summer heat load.

36. Properly lubricate all pneumatic components such as cylinders, rotating compressed air tools, valves, etc.

Building Information for Total Energy Management

Appendix B

Surveyed by: _____
Survey Date: _____

I. GENERAL INFORMATION

Identity:

Name of building _____

Address _____

_____

Type(s) of occupancy _____

_____

Name of owner(s) _____

Person(s) in charge of building _____

_____

Physical data:

Building orientation _____

No. of floors _____

Floor area, gross, square feet _____

Net air conditioned square feet _____

Construction type:

    Walls (masonry, curtain, frame, etc.)

    N _____ S _____ E _____ W _____

  Roof:

    Type: Flat _____   Color: Light _____

             Pitched _____   Dark _____

  Glazing:

Exposure	Type*	%Glass/Exterior wall area
N	_____	_____
S	_____	_____
E	_____	_____
W	_____	_____

*Type: Single, double, insulating, reflective, etc.

Glass shading employed outside (check one):
Fins _____ Overhead _____ None _____ Other _____
Glass shading employed inside (check one):
Shades ___ Blinds ___ Drapes, open mesh ___ Drapes opaque ___ None ___ Other

Sketch of building showing principle dimensions:

Building type:
All electric _____
Gas total energy _____
Oil total energy _____
Other _____

BUILDING OCCUPANCY AND USE:
Weekdays: Occupied by:* _____ people from _____ to _____ (hours)
                           _____   _____   _____
                           _____   _____   _____
                           _____   _____   _____
                           _____   _____   _____

Saturdays: _____   _____   _____
Sundays, holidays   _____   _____   _____

Hours air conditioned: Weekdays from _____ to _____; Saturdays _____ to _____
Sundays, holidays from _____ to _____
*(Account for 24 hours a day. If unoccupied, put in zero)

II. ENVIRONMENTAL CONDITIONS

Outdoor conditions:

Winter: Day _____ °F dB _____ mph wind
              _____ °C dB _____ kmh wind
Summer: Day _____ °F dB _____ mph wind
              _____ °C dB _____ kmh wind

Night _____ °F dB _____ mph wind
      _____ °C dB _____ kmh wind
Night _____ °F dB _____ mph wind
      _____ °C dB _____ kmh wind

Maintained indoor conditions:

Winter: Day _____ °F dB _____ %rh
              _____ °C dB
Summer: Day _____ °F dB _____ %rh
              _____ °C dB

Night _____ °F dB _____ %rh
      _____ °C dB
Night _____ °F dB _____ %rh
      _____ °C dB

III. SYSTEMS AND EQUIPMENT DATA

HVAC systems:

Air-handling systems (check as appropriate):

Perimeter system designation:
- Single zone _____ Multizone _____
- Fan coil _____ Induction _____
- Variable air volume _____ Dual duct _____
- Terminal reheat _____ Self-contained _____
- Heat pump _____

Interior system designation:
- Fan coil _____ Variable air volume _____
- Single zone _____ Other (describe) _____

Principle of operation:

Perimeter:
- Heating-cooling-off _____
- Air volume variation _____
- Air mixing control _____
- Temperature variation _____

Interior:
- Heating-cooling-off _____
- Air volume variation _____
- Temperature variation _____

All fans (supply, return and exhaust):

Location	Horsepower	Type	Method of Operation
_____	_____	_____	_____
_____	_____	_____	_____
_____	_____	_____	_____

Source of heating energy:
- Hot water _____ Steam _____ Electric resistance _____ Other _____

Heating plant:
- Boiler No. _____ Rating _____ MBH
- _____ _____

Boiler type:
- Fire tube _____ Water tube _____ Elec. resist. _____ Electrode _____ Other _____
- Fuel used _____ Standby _____
- Hot water supply _____ °F, Return _____ °F
- _____ °C _____ °C

Steam pressure _____ psi

Pumps No. _____ Total hp _____

Room heating units:
- Type: Baseboard _____ Convectors _____ Fin tube _____
- Ceiling or wall panels _____ Unit heaters _____ Other _____

248   Energy Conservation in Buildings and Industrial Plants

Cooling plant:
   Chillers: No. _____ Total capacity (tons) _____
   Type: Centrifugal _____ Reciprocating _____ Absorption _____
   Capacity controlled by: _____
   Chiller operation: Starting controls _____
                        Stopping controls _____
   Chilled water temp. supply _____ °F, return _____ °F
                              _____ °C   _____ °C
   Condenser water temp. _____ in °F _____ out °F
                              _____ in °C _____ out °C
Heat dissipation device:
   Evaporative condenser _____
   Air-cooled condenser _____
   Cooling tower _____
Condenser/cooling tower fan hp _____
Heat recovery device: Double-bundle condenser _____ Other _____
Chilled water pumps _____ Total hp _____
Condenser water pumps _____ Total hp _____
Self-contained units:
   Type: Through-the-wall-air conditioner _____ Other _____
   No. of units _____ Basic module served _____
   Capacity (tons) _____

Energy conservation devices:
   Type:
      Condenser water used for heating _____
      Demand limiters _____
      Energy storage _____
      Heat recovery wheels _____
      Enthalpy control of supply-return-exhaust damper _____
      Recuperators _____
      Others _____

Lighting:
   Interior lighting type: _____
   W/ft^2:   Hallway/corridor _____
             Work stations _____
             Circulation areas within work space _____
   On/off from breaker panel _____ Wall switches _____
   Control switching _____
   Exterior Lighting: Type _____ Total kW _____

Domestic hot water heating:
   Size _____ Rated input _____
   Energy Source: Gas _____ , Oil _____ , Electric _____ , Other _____

Other equipment (Elevators, Data Processing, Kitchen, etc.):

Equip. Description	Quantity	Size/Capacity in Btu (kJ), kW, hp, etc.
_____	_____	_____
_____	_____	_____
_____	_____	_____
_____	_____	_____

IV. **OPERATING SCHEDULE:**
Operation (Start/stop)

Equipment description	Weekdays	Saturday	Sunday	Holiday
Refrigeration cycle mach.	_____	_____	_____	_____
Fans, supply	_____	_____	_____	_____
Fans, return/exhaust	_____	_____	_____	_____
Fans, exhaust only	_____	_____	_____	_____
HVAC auxiliary equip.	_____	_____	_____	_____
Lighting, interior	_____	_____	_____	_____
Lighting, exterior	_____	_____	_____	_____
Elevators	_____	_____	_____	_____
Escalators	_____	_____	_____	_____
Domestic hot water ht.	_____	_____	_____	_____
Other (describe: _____)	_____	_____	_____	_____

SOURCE: "Total Energy Management: A Practical Handbook on Energy Conservation and Management," National Technical Information Service, PB-265-683, March 1976.

Conversion and Other Useful Tables

Appendix C

Approximate Conversion Factors for Crude Oil*

From \ Into	Metric Tons	Long Tons	Short Tons	Barrels
		Multiply By		
Metric tons	1	0.984	1.102	7.33
Long tons	1.016	1	1.120	7.45
Short tons	0.907	0.893	1	6.65
Barrels	0.136	0.134	0.150	1
Kiloliters (cubic meters)	0.863	0.849	0.951	6.29
1000 gal (Imp.)	3.91	3.83	4.29	28.6
1000 gal (U.S.)	3.25	3.19	3.58	23.8

	Kiloliters (Cubic Meters)	1000 gal (Imp.)	1000 gal (U.S.)
		Multiply By	
Metric tons	.116	0.256	0.308
Long tons	1.18	0.261	0.313
Short tons	1.05	0.233	0.279
Barrels	0.195	0.035	0.042
Kiloliters (cubic meters)	1	0.220	0.264
1000 gal (Imp.)	4.55	1	1.201
1000 gal (U.S.)	3.79	0.833	1

*Based on world average gravity (excluding natural gas liquids).

Btu Heat Values*

	Crude Petroleum, 42-gal Barrel)	Anthracite Coal, Short Ton	Bituminous Coal, Short Ton	Natural Gas—Dry, 1000 ft³	Distillate Fuel Oil 42-gal Barrel	Residual Fuel Oil, 42-gal Barrel	Liquefied Pet. Gas, 42-gal Barrel
Crude petroleum, 42-gal barrel equals		0.228	0.221	5.604	0.996	0.923	1.446
Anthracite coal, short ton equals	4.379		0.969	24.541	4.361	4.040	6.333
Bituminous coal and lignite, short ton equals	4.517	1.031		25.314	4.498	4.167	6.532
Natural gas—dry, 1000 ft³ equals	0.178	0.041	0.040		0.178	0.165	0.258
Distillate fuel oil, 42-gal barrel equals	1.004	0.229	0.222	5.628		0.927	1.452
Residual fuel oil, 42-gal barrel equals	1.084	0.248	0.240	6.074	1.079		1.567
Liquefied pet. gas, 42-gal barrel equals	0.692	0.158	0.153	3.875	0.689	0.638	
Btu heat values as used	5800	25,400	26,200	1035	5825	6287	4011

*Other refined products Btu values (1000s): gasoline 5248; kerosene 5670; lubricants 6064.8; wax 5537.3; asphalt 6636; miscellaneous 5796; natural gasoline 4620/42-gal barrel.
SOURCE: *Annual Statistical Review, Petroleum Industry Statistics, 1964–1973*, API, September 1974.

Project Year Discount Factors—Single Amount

Present value of $1 (single amount—to be used when cash flows accrue in different amounts each year)

Project Year	Discount Rate				
	6%	8%	10%	12%	15%
1	0.972	0.963	0.954	0.946	0.935
2	0.917	0.892	0.867	0.845	0.813
3	0.865	0.826	0.788	0.754	0.707
4	0.816	0.764	0.717	0.674	0.615
5	0.770	0.708	0.652	0.602	0.535
6	0.726	0.655	0.592	0.537	0.465
7	0.685	0.607	0.538	0.480	0.404
8	0.627	0.562	0.489	0.428	0.351
9	0.610	0.520	0.445	0.382	0.306
10	0.575	0.482	0.405	0.314	0.266
11	0.543	0.446	0.368	0.304	0.231
12	0.512	0.413	0.334	0.272	0.201
13	0.483	0.382	0.304	0.243	0.175
14	0.456	0.354	0.276	0.217	0.152
15	0.430	0.328	0.251	0.194	0.132
16	0.410	0.304	0.228	0.173	0.115
17	0.382	0.281	0.208	0.154	0.100
18	0.361	0.260	0.189	0.138	0.087
19	0.340	0.241	0.172	0.123	0.076
20	0.321	0.223	0.156	0.110	0.066
21	0.303	0.207	0.142	0.098	0.057
22	0.286	0.191	0.129	0.088	0.050
23	0.270	0.177	0.117	0.078	0.043
24	0.254	0.164	0.107	0.070	0.038
25	0.240	0.152	0.097	0.062	0.033
26	0.226	0.141	0.088	0.056	0.028
27	0.214	0.130	0.080	0.050	0.025
28	0.202	0.121	0.073	0.044	0.022
29	0.190	0.112	0.066	0.040	0.019
30	0.179	0.103	0.060	0.035	0.016
35*	0.152	0.084	0.047	0.026	0.011
40*	0.114	0.057	0.029	0.015	0.006
45*	0.085	0.039	0.026	0.008	0.003
50*	0.064	0.026	0.017	0.005	0.001

Note: Table factors represent an arithmetic average of beginning and end of the year single-amount factors found in standard present value tables:
Formula: Discount factor = $1(1 + r)^n$, where r = discount rate as decimal, n = project year.
*Values for project years 35 through 50 are arithmetic average of beginning and end of the 5-year interval.

EXAMPLES OF DEPRECIATION

1. Total depreciable amount, $1000 (no salvage).
2. Life of project, $n = 5$ years.
3. For sum-of-the-years digits, the fraction denominator is $1 + 2 + 3 + 4 + 5$, or $5(5 + 1)/2 = 15$.

	Project Year					
	1	2	3	4	5	Totals
Straight line (20%/yr):						
Annual depreciation: $1000/5 yr	$200	$200	$200	$200	$200	$1000
Depreciated value	$800	$600	$400	$200	0	
Sum-of-years digits:						
Numerator	5	4	3	2	1	
Fraction	5/15	4/15	3/15	2/15	1/15	
Annual depreciation	$333	$267	$200	$133	$ 67	$1000
Depreciated value	$667	$400	$200	$ 67	0	
Declining balance at twice straight line (40%/yr):						
Annual depreciation	$400	$240	$144	$ 86	$ 52	$ 922
Depreciated value	$600	$360	$216	$130	$ 78	

SOURCE: "Guidelines for Energy Conservation in Industrial Processes," prepared for Southern California Gas Company by Envirodyne Energy Services Inc., April 1974.

EXAMPLE OF CASH FLOW ANALYSIS

1. Initial installation cost, $1000.
2. No salvage value.
3. Project life, 5 years.
4. Total annual operating savings, $400.
5. Depreciation basis for income tax, sum-of-the-years' digits.
6. Net operating savings (for income tax): Total operating savings less depreciation.
7. Income tax, 50 percent of net operating savings.
8. Net cash flow (after initial installation): Total operating savings less income tax.
9. Negative values shown in parentheses.

Project Year	Total Oper. Savings	Depreciation	Net Oper. Savings	Income Tax	Net Cash Flow
0	0	0	0	0	(1000)
1	400	(333)	67	(33)	367
2	400	(267)	133	(67)	333
3	400	(200)	200	(100)	300
4	400	(133)	267	(133)	267
5	400	(67)	333	(167)	233
				NET	500

SOURCE: "Guidelines for Energy Conservation in Industrial Processes," prepared for Southern California Gas Company by Envirodyne Energy Services Inc., April 1974.

COMPARATIVE ENERGY COSTS BY STATE

Fuel oil is more than twice as expensive as coal as an energy source, and electricity is more than three times the cost of oil, according to these figures, given in dollars per million British thermal units.

The aim is to give 1977 costs by state for each fuel, using the available data most appropriate for large industrial plants. For coal, the 12 months ended October 31 are used, to reduce price distortions caused by the coal strike and anticipations of it.

Fossil fuel costs are those reported by each state's electric utilities to the Federal Energy Regulatory Commission. Electricity costs are based on the revenue and sales by investor-owned utilities to large industrial and commercial customers, as reported to the Edison Electric Institute.

The national averages, in customary units, come to $13.69/bbl for oil, $1.34/1000 ft^3 for gas, $19.84/ton for coal, and 2.52 cents/kWh for electricity.

In a few cases, substitute figures are provided. These are footnoted.

	Fuel Oil	Natural Gas	Coal	Electricity
Alabama	$2.78	$1.75	$1.11	$7.77
Arizona	2.31	1.09	0.43	9.44
Arkansas	2.03	0.97		6.89
California	2.37	2.10		9.44
Colorado	1.94	1.06	0.60	5.83
Connecticut	2.24	2.71*		10.29
Delaware	2.21	1.84	1.06	9.32
Dist. of Columbia	2.25	1.58*		8.03
Florida	2.13	0.88	1.23	8.76
Georgia	2.38	1.30	1.08	7.42

	Fuel Oil	Natural Gas	Coal	Electricity
Idaho	$2.89†	$1.94†		$3.40
Illinois	2.50	1.88	$0.98	7.21
Indiana	2.28	1.41	0.79	7.21
Iowa	2.66	1.30	1.00	7.36
Kansas	2.10	1.01	0.74	7.42
Kentucky	2.86	0.98	0.81	5.39
Louisiana	1.96	0.91		4.48
Maine	2.05	2.71*		6.27
Maryland	2.15	2.08†	1.19	8.03
Massachusetts	2.12	1.88		11.84
Michigan	2.64	1.85	1.08	8.65
Minnesota	2.09	1.11	0.73	8.29
Mississippi	1.85	1.44	1.07	8.76
Missouri	2.39	1.12	0.72	7.21
Montana	2.02	0.67	0.28	3.28
Nebraska	1.81	0.97	1.04	
Nevada	2.16	1.74	0.55	8.73
New Hampshire	2.06	2.71*	1.28	9.23
New Jersey	2.32	1.79	1.41	11.14
New Mexico	2.49	1.44	0.28	7.42
New York	2.19	1.40	1.16	10.14
North Carolina	2.72	1.58*	1.20	6.80
North Dakota	2.76	1.23	0.37	8.59
Ohio	2.54	1.38	0.99	5.92
Oklahoma	2.67	1.30	1.02	6.57
Oregon	2.54	1.93		3.81
Pennsylvania	2.35	2.07	1.00	8.47
Rhode Island	2.07	2.71		11.61
South Carolina	1.84	1.58*	1.18	6.39
South Dakota	2.15	1.20	0.55	6.86
Tennessee	3.01	1.39	1.00	6.92
Texas	2.10	1.23	0.56	6.10
Utah	2.12	1.09	0.71	6.48
Vermont	2.81	2.01	1.32	7.65
Virginia	2.16	1.83	1.31	8.00
Washington		2.03*	0.73	3.43
West Virginia	2.80	1.97	1.00	6.42
Wisconsin	2.60	1.82	0.95	7.15
Wyoming	3.40	2.08	0.34	3.81
Nationwide	2.23	1.30	0.92	7.39

*Based on gas sales to the industrial sector by distributors reporting to the American Gas Association. These are the appropriate regional figures, not costs for the one state alone.
†Based on electric utilities' purchases at facilities other than steam-electric ones.
SOURCE: *Energy User News* (Sept. 4, 1978).

Glossary

Absorption A process whereby heat extracts one or more substances present in an atmosphere or mixture of gases or liquids, accompanied by physical change or chemical changes.

Absorption chiller A refrigeration unit based upon absorption refrigeration.

Absorption refrigeration Cooling caused by the expansion of liquid ammonia into gas and water. Heat is the primary source of energy!

ACS Automatic control systems.

Active energy systems When applied to building design, it refers to the mechanical transfer of energy by the use of an external energy source.

Address A coded representation of the origin or destination of a computer data message.

AEB Annual energy budget.

Air change The movement of a volume of air in a given period of time in or out of a building or room. Air changes are expressed in cubic feet per minute.

Alphanumeric Alphabetic characters, numeral or specific symbols.

Ambient temperature The temperature of the outside or surrounding air.

Analog The representation of numerical quantities by varying the amplitude of control signals (current, voltage, or resistance).

Annunciator A device for calling the operator's attention to the status of a point.

ASHRAE American Society of Heating, Refrigerating, and Air Conditioning Engineers.

Baud A unit of signaling speed equal to the number of signals per second, same as bits per second when a bit constitutes a signal.

Binary point (sensor) A sensor whose output indicates that either a monitored characteristic exists or it is zero. For example, an air vane in a fan airstream indicates that the fan is on or it isn't.

Bit A contraction of binary digit, either of the characters 0 or 1.

Boiler A device used to heat water and/or produce steam. Major components include burner, heat exchanger, flue and expansion chamber, and controls. Traditional fuels are oil, gas, coal, and electricity.

Boiler efficiency The ability of a unit to convert a form of energy (gas, oil, coal, electricity) to heat energy at the highest possible rate; that is, 80–85 percent efficiency should be an attainable goal.

BOMA Building Owners and Managers Association.

British thermal unit The amount of heat necessary to raise the temperature of one pound of water one degree Fahrenheit. A unit of thermal (heat) energy approximately equal to the amount of heat given off by burning a kitchen match.

Btu (See *British thermal unit*.)

Building skin The physical building elements which envelop a structure.

Building orientation Considered in relation to the on-site orientation, it is concerned with the direction of prevailing winds and position of north-south, east-west exposures as these relationships directly impact on heat loss-gain properties of a building.

Bypass loop A piping configuration which bypasses or circles the flow of a heat-absorbing medium around rather than through a piece of mechanical equipment.

Card A hard-wired piece of equipment located in the remote panel that is specially wired for remote-point functions.

Capacity The usable output of a system or system component.

Cathode-ray tube Electronic viewing screen used for display of data and information.

Central building automation system A system that monitors and controls remote mechanical, electrical, utility, and life safety building systems from a central location.

Central processing unit That part of a computer that contains the main core storage and arithmetic units which control and perform the execution of instruction.

Centrifugal fan A device for propelling air by centrifugal action.

Chimney effect The tendency of air or gas in a duct or other vertical passage to rise when heated due to its lower density compared with that of the surrounding air. In buildings, the tendency toward displacement (caused by the difference in temperature) of internal heated air by unheated outside air due to the difference in density of inside and outside air.

Coefficient of utilization The ratio of lumens on a work plane to lumens emitted by the lamps.

Comfort zone The ranges of indoor temperature, humidity, and air movement under which most persons enjoy mental and physical well-being.

Compression refrigeration system A process by which the pressure and temperature of the refrigerant is increased to allow for greater heat transfer.

Conduction A process of heat transfer whereby heat is transmitted through a material.

Controller A device that measures changes in controlled variables in rooms, ducts, and liquids and sends an appropriate signal to adjust such system functions.

Control set-point adjustment A method by which the analog set point of a controller is changed to a new value.

Convection Transfer of thermal energy (heat) by the movement of a fluid or gas.

CPU (See *Central processing unit.*)

CRT (See *Cathode-ray tube.*)

CU (See *Coefficient of utilization.*)

Damper A device used to vary the volume of air passing through an air outlet, inlet, or duct.

Degree-day The difference between 65°F and the average daily temperature; i.e., average daily temperature = 25°F; therefore, the degree-day would be 65 − 25 = 40. The greater the number of heating or cooling degree-days, the higher the energy consumption.

Demand factor The ratio of the maximum demand of a system, or part of a system, to the total connected load of a system, or part of a system, under consideration.

Dewpoint The temperature at which water vapor begins to condense.

Digital Data representing a finite quantity or condition by the number of discrete, identical pulse signals per time interval.

Disc A nonprogrammable bulk storage, random-access memory consisting of magnetizable coating on one or both faces of a coated, thin circular plate.

Discrete Synonymous with digital. Can also mean data describing the status of a two-position control point.

DOE Department of Energy.

ECM Energy conservation measure.

Economizer cycle A method of operating a ventilation system to reduce refrigeration load. Whenever the outdoor-air conditions are more favorable (lower heat content) than return-air conditions, outdoor-air quantity is increased.

ECS (See *Energy control system.*)

EEI Edison Electric Institute.

Energy control system A system designed and operated to control the energy-consuming equipment of an institution or installation of buildings, usually automatically, to optimize energy conservation.

Efficacy of fixtures Ratio of usable light to energy input for a lighting fixture or system (lumens/watt).

EMCS Energy management control system.

Energy The capacity for doing work; taking a number of forms which may be transformed from one into another, such as thermal, (heat), mechanical (work), electrical, and chemical; in customary units, measured in kilowatthours (kWh) or British thermal units (Btu); in SI units, measured in joules (J) where 1 joule = 1 wattsecond.

Energy audit Examination of a building's "as built" architectural and mechanical drawings, energy usage patterns, and fuel history to identify energy-conserving opportunities.

Energy cost savings Energy savings times energy unit (therm, 10^3 ft^3, kWh) Note: The results indicate a *cost avoidance* based upon anticipated reduction in energy consumption.

Enthalpy For the purpose of air conditioning, enthalpy is the total heat content of air above a datum usually in units of Btu/lb. It is the sum of sensible heat and latent heat and ignores internal energy changes due to pressure change.

EPRI Electric Power Research Institute.

Evaporative cooling Lowering the temperature of a large mass of liquid by utilizing the latent heat of vaporization of a portion of the liquid cooling air by evaporating water into it.

Executive program That portion of a computer program that manages in proper sequence the execution of the subroutines and other portions of the program and allocates resources of the computer system, such as timekeeping, printing output, etc.

FEA Federal Energy Administration.

First cost Original construction cost.

Flow rate Velocity at which a fluid travels, usually through an opening or duct.

Flue gas analysis A test procedure whereby the relationship between air, fuel, and stack temperatures are monitored, thus indicating the apparent transfer of energy during combustion of the fuel.

Footcandle Energy of light at a distance of one foot from a standard (sperm oil candle).

Format The predetermined arrangement of field, lines, page numbers, functions, and similar characters of written data.

Greenhouse effect The effect of the earth's atmosphere in trapping heat from the sun. The atmosphere acts like a greenhouse by admitting the sun's shortwave radiation but blocking the exit of long-wavelength radiation reemitted by the earth or other warmed objects.

Hardware The magnetic, electrical, electronic, or mechanical devices or components of a system or computer.

Heat, latent The quantity of heat required to cause a change in state.

Heat, sensible Heat that results in a temperature change but no change in state.

Heat capacity Sometimes called the thermal capacity, a measure of how much heat is required to raise the temperature of a specific quantity of given material by a given amount.

Heat exchanger A device used to transfer heat from one medium to another.

Heat gain As applied to HVAC calculations, it is that amount of heat gained by a space from all sources, including people, lights, machines, sunshine, etc. The total heat gain is the quantity of heat that must be removed from a space to maintain indoor comfort conditions.

Glossary

Heat loss A decrease in the amount of heat contained in a space, resulting from heat flow through walls, windows, roof, and other building envelope components.

Heat pump A reversible refrigeration system that delivers more heat energy to the end use than is put in the compressor. The additional energy input results from the absorption of heat from a low-temperature source—the combination heating and cooling unit. It operates like a normal air conditioner in summer, and in winter operates in reverse, ejecting warm air indoors and cool air (or water) outdoors.

Heat wheel A device used in ventilating systems which tends to bring incoming air into thermal equilibrium with exiting air. As a result, hot summer air is cooled and winter air is warmed.

HVAC Heating, ventilating, and air conditioning systems.

IES Illuminating Engineering Society.

IGT Institute of Gas Technology.

Infiltration The process by which outdoor air leaks into a building by natural forces, especially as through cracks around doors and windows, etc. (usually undesirable).

Insolation The amount of solar radiation on a given plane. Expressed in langleys, or Btu/ft².

Insulation thermal Any material high in resistance to heat transmission that when placed in the walls, ceilings, or floors of a structure will reduce the rate of heat flow.

Interface A common boundary between automatic data processing systems and parts of a single system.

Kilowatthour Unit of electrical energy consumption which equals about 3400 Btu.

kWh (See *kilowatthour*.)

Langley Measurement of radiation intensity. One Langley = 3.68 Btu/ft².

Life-cycle cost The cost of equipment over its entire life including operating and maintenance costs.

Line printer A printer device that can simultaneously print several character graphics as a permanent record.

Load Leveling Deferment of specified loads to limit electrical power demand to a predetermined level.

Load profile Time distribution of building heating, cooling, and electrical load.

Lumen Unit of luminous flux.

Luminaire Light fixture designed to produce a specific effect.

Masonry Stone, brick, concrete, hollow tile, concrete block, gypsum block, or other similar building units or materials, or combination of the same, bonded together with mortar to form a wall, pier, buttress, or similar mass.

Mass The property of density of a material. The use of mass by itself or in combination with insulation gives building structures the capacity to store thermal energy.

MCC (See *Motor control center.*)

Microcomputer A hard-wired logic system that is physically wired to perform specified functions. Rewiring is required to change existing functions or to add additional functions.

Microprocessor A large-scale integration processing unit containing a single integrated circuit (IC) chip or a set of IC chips with limited memory.

Minicomputer A small computer with shorter word lengths, smaller size, limited processing capabilities, and lower prices, distinguished from microcomputer by CPU. The latter has large-scale integration CPU.

MIUS Modular integrated utility systems.

Modem (modulator-demodulator) Converts outgoing pulses to a continuous, modulated audio signal that will not be degraded when transmitted over a telephone line, and reconverts incoming modulated signals back into pulses. The reference characteristic of a modem is its spread or data rate, expressed in bits per second (bps), which is also known as baud.

Moisture content The relative quantity of water in a volume of air expressed in grains of moisture; a grain of moisture equals approximately $1/7000$ lb.

Motor control center Enclosure for accommodating motor starters and related controls and accessories.

Multiplex (matrix) Transmission of multiple signals over a group of conductors which are arranged by rows and columns to form a matrix.

Multiplex (time-shared) Transmission of multiple signals over the same path by using different time intervals for each transmission.

NEMA National Electric Manufacturers Association.

Orsat apparatus A device for measuring the combustion components of boiler or furnace flue gases.

OSES On-site energy system.

Pay back period The time it requires for anticipated energy savings to recover the cost of the investment; also termed recovery rate.

Peak load Maximum predicted energy demand of a system.

PEDS Peltier-effect diffusion system (or still).

Peripheral Input-output equipment used to communicate with the CPU, typewriters, printers, and CRT display.

Piggyback operation Arrangement of chilled water generation equipment which permits exhaust steam from a steam turbine-driven centrifugal chiller to be used as the heat source for an absorption chiller.

Point The address of a field input-output device which may be discrete or analog.

Point (analog) An address having a continuous, variable control signal representing values such as 456 psig, 60 gal/min, 72°F, and 20 mA.

Point (discrete) An address having a discrete value such as on, off, opened, and closed.

Power factor The relationship between kVA and kW. When the power factor is unity, kVA equals kW.

Program A series of instructions which define in detail the computer steps necessary to perform a function.

Psychrometric Pertaining to the device comprising two thermometers, one a dry bulb, the other a wet bulb or wick-covered bulb, used in determining the moisture content or relative humidity of air or other gases.

Reflectance A property of a material indicating the percentage of light that is reflected when a certain amount of light strikes the surface of the material or is transmitted through it.

Relative humidity The ratio of the amount of water vapor at a given temperature to the maximum amount of water vapor that could be held as vapor.

Relay A device for converting an electrical or pneumatic signal into an electromagnetic switching device having electrical contractors energized by electrical current through its coil.

Remote terminal panel The remote field wiring cabinet to which all input- and output-controlled devices and sensors connect.

Retrofit Alteration or modification of an existing structure, or mechanical system or components therein to conserve energy.

Set back The lowering of the thermostat setting. The technique is used to reduce the amount of energy required to heat a space.

Shading coefficient The effectiveness of any shading device can be expressed by its shading coefficient which describes the fraction of the incident solar energy transmitted through a window.

Shortwave radiation A term used loosely to distinguish radiation in the visible and near-visible portions of the electromagnetic spectrum (roughly 0.4 to 1.0 μm in wavelength) from longwave radiation (infrared radiation).

Sling psychrometer A psychrometer in which the wet and dry bulb thermometers are mounted upon a frame and connected to a handle at one end by means of a bearing or a length of chain. It may be whirled in the air in such a manner to provide the simultaneous measurement of wet and dry bulb temperatures.

Software The internal programs or routines prepared to simplify programming and computer operations. Such routines permit the operator to use English or mathematics to communicate with the computer; various programming aids supplied by manufacturers to facilitate operation of the equipment, such as various assemblers, subroutine libraries, compilers, operating systems, and industry application programs.

Solar intensity The amount of solar radiation.

Sol-air temperature   The theoretical air temperature that would give a heat flow rate through a building surface equal in magnitude to that obtained by the addition of conduction and radiation effects.

Solstice   The two days (actually instants) during the year when the earth is so located in its orbit that the inclination (about 23½°) of the polar axis is toward the sun. The days are June 21 for the North Pole and December 22 for the South Pole; because of leap years, the dates vary slightly.

Space heating   Interior heating of a building or room.

TES   Thermal energy storage.

Thermal lag   The ability of materials to delay the transmission of heat; can be used interchangeably with time lag.

Ton of air conditioning   The thermal refrigeration energy required to melt one ton of ice (2000 pounds). Also a means of expressing cooling capacity. (1 ton = 12,000 Btu/h cooling.)

Truck wiring (matrix)   The multipair cable or cables connected between the central processing unit (CPU) and the remote terminal panels (RTP).

Truck wiring (time-shared)   The transmission line used to send digital signals between the CPU and RTP.

U value   A coefficient which indicates the energy (Btu) which passes through a component for every degree Fahrenheit of temperature difference from one side to the other under steady-state conditions.

Vapor barrier   A component of construction impervious to the flow of moisture or air.

Vapor pressure   The force exerted when moist air attempts to seek equilibrium by migrating to areas with less water vapor.

Ventilation   Air introduced to an occupied space or building to accomplish comfort and odor control. Commonly refers to outside-air quantities mixed with room air.

Weatherstripping   Foam, metal, or rubber strips used to form a seal around windows, doors, or openings to reduce infiltration.

Wet bulb temperature   The lowest temperature attainable by evaporating water into the air without the addition or subtraction of energy.

Index

ACS (automatic control systems), 74–75
AEB (annual energy budgets), 24–26
Air-supported structures, 18
Air-to-air heat exchangers, 39
Algorithms in energy conservation programs, 95–99
American Paper Institute, 165
American Society of Heating, Refrigeration, and Air Conditioning Engineers (ASHRAE), 33
Annual energy budgets (AEB), 24–26
ASHRAE Standard 90-75, 33–35, 63
Automated energy management systems, 64–67
   controls, 73–75
      central automatic, 73
      central manual, 73
      local, 73
Automatic control systems (ACS), 74–75
Automatic lighting control, 31–33
Automation, levels of, 89–90

Balance of payments, U.S., 219, 227
Borek, G. S., 37
Bubble buildings, 18
Building materials:
   concrete, 2, 3
   energy intensiveness of, 2
   fluid-filled, 18–19
   glass, 6
   plastics, 2
   thermal inertia of, 9–11
   (See also Cement industry; Iron and steel industry; Pulp and paper industry)
Building orientation, 6
Building selection factors for energy conservation, 82–85
Building types:
   air-supported structures, 18
   bubble buildings, 18
   hydraulic structures, 17–18
   pneumatic structures, 17–18
   self-built structures, 17

Buildings, energy flow and conversion in, 62–64
Bureau of Mines, U.S., 222

Carbon dioxide, 209–213
Carbon monoxide, 203
Carter, Jimmy, administration of, 226
Cash flow analysis, 192, 254
Cement industry, 2, 3, 162, 170–171
Chemical energy, direct conversion of, 174–176
Chemical industry, 132, 162, 171–176
Chilled water computer optimization program, 75–78
Chiller double-bundle condensers, 46, 138
Chiller-loading computer optimization program, 81–82
Chiller systems, 46–48
Clean Air Acts (1976 and 1977), Amendments, 200, 213
Coal gasification, 197
Coal reserves, U.S., 201, 223
Cogeneration systems, 115, 128–132
   combined cycle, 131, 133
   diesel topping cycle, 131
   gas turbine topping cycle, 131
   organic bottoming cycle, 132, 133
   steam bottoming cycle, 132, 133
   steam topping cycle, 131
Columns, prestressed, 18–19
Combined-cycle generation, 115, 131, 133, 143
Combined-cycle power plant, 144
Combustion process:
   adverse weather effects from, 209, 211–213
   combustion equipment checklist, 242
   (See also Fossil-fueled electric plants)
Computer modeling, 90–95
Computer optimization programs, 75–82
   chilled water optimization programs, 75–78

266 Index

Computer optimization programs *(Cont.):*
  chiller-loading optimization programs, 81–82
  electrical demand load leveling, 78–79, 107
  electrical demand load limiting, 79
  electrical demand load profiles, 80
  electrical demand load shedding, 65, 79–80, 104, 112
  enthalpy optimization, 75–77
  power-factor correction, 60–61, 81
Computerized energy management systems, 70–73
Computerized power management systems, 69
Concrete, 2, 3
  (*See also* Cement industry)
Condensers, double-bundle, 46, 138
Congress, U.S., 226, 234
Construction industry, 6
Construction materials (*see* Building materials)
Control-point combinations, heating and cooling system, 85–88
Controls for automated energy management systems, 73–75
Conversion tables, 251–253
Cross-media environmental impacts, 216, 231

Demand-limit controllers, 60
Demand limiting, 59–60
Department of Commerce, U.S., 227
Department of Energy (DOE), U.S., 229–230
Depreciation, 254, 255
Designing with nature, 12
Diesel generators, 124, 131
Direct conversion of chemical energy, 174–176
Distillation processes, 177–185
Dollar, U.S., value of, 220
Double-bundle condensers, 46, 138
Dual fuel turbines, 124

ECM (energy conservation measures), 28, 37, 60, 67, 170
Economic analysis of on-site energy generation systems, 125–128
ECS (energy conservation studies), 23–24
Edison Electric Institute (EEI), 206
Electric plants, fossil-fueled, 197–202
Electric Power Research Institute (EPRI), 197, 200

Electric rate structure:
  rate incentives and disincentives, 103
  declining block rate structure, 110
  peak-period pricing, 110
  seasonal rates, 110
  time-of-day rates, 110–113
Electric utility industry, 103–104
Electrical demand limiting, 59–60
Electrical demand load leveling, 78–79, 107
Electrical demand load profiles, 80
Electrical demand load shedding, 65, 79–80, 104, 112
Electrical designer, role of, 31–33
EMCS (energy management control systems), 83–85
Energy audits, 229, 237
Energy conservation checklist, 237–244
Energy conservation measures (ECM), 28, 37, 60, 67, 170
Energy conservation programs, 48–51, 95–99
  building selection factors for, 82–85
Energy conservation studies (ECS), 23–24
Energy consumption forecasts, industrial, 185–194, 206
  and conservation, 194–195, 242–244
Energy costs, comparative, by state, 255–256
Energy flow and conversion in buildings, 62–64
Energy flow sheets, 26–28
Energy forecasting scenarios, 190–194
Energy generation systems, economic analysis of, 117, 125–128, 212
Energy-intensive industries, 161
  (*See also specific industry*)
Energy intensiveness of building materials, 2
Energy management control systems (EMCS), 83–85
Energy management systems:
  automated, 64–67
  computerized, 70–73
  controls, 73
  optimizing packages for, 70–71
Energy planning, 185–194, 206
Energy policy, U.S., 219–236
Energy storage, thermal, 11, 15–16, 106–109
Energy systems, on-site (*see* On-site energy systems)
Energy Tax Act (1978), 229–231
Energy technology, environmental impacts of, 197–217
  cross-media, 216, 231
Energy wastefulness, 19, 53, 62

Engines:
  Rankine-cycle, 139–142
  reciprocating, 124–125
Enthalpy optimization, 75–77
Environmental impacts of energy technology, 197–217
  cross-media, 216, 231
Environmental Protection Agency (EPA), U.S., 200–201, 213, 229, 232
Environmental regulations:
  federal air pollution laws, 230
    expenditures for compliance with, 214
  siting laws, 206–209
  state air pollution laws, 230
  state environmental laws, 206–209
EPRI (Electric Power Research Institute), 197, 200
Evaporation processes, 177–185

Fairbanks, R. F., 37
Federal air pollution laws, 230
  expenditures for compliance with, 214
Federal Energy Administration (FEA), 29
Federal Energy Regulatory Commission, 231
Federal Power Commission, 206
Fenestration, 32
Fluid-filled building elements, 18–19
Flywheels:
  with Rankine-cycle engines, 138
  in thermal energy storage applications, 109
Footswitch controls, 62
Forecasting in energy planning, 185–194, 206
  and conservation, 194–195, 242–244
Forest Service, U.S., 1
Fossil-fueled electric plants, 197–202
  (See also Combustion process)
Franssen, H. T., 226–227
Fuel cells, 174–176
Fuel reserves, 222–224
Fuel resources, 222–229
Fuel substitution, intraindustry, 157–160, 177, 194

Gambs, Gerard C., 1
Gas, liquified, 225
Gas turbines, 124, 131
General Services Administration (GSA), 29
Generators:
  diesel, 124
  induction motor-generators, 137

Geological Survey, U.S. (USGS), 222
Geothermal energy, 143–147
Glass, 6
Greenhouse effect, 209–212
GSA (General Services Administration), 29

Hays, Dennis, 103
Heat conduction, 7
Heat convection, 7
Heat exchangers, air-to-air, 39
Heat pump transfer systems, 42–45
Heat pumps, 42–46
  air-source, 43–44
  air-to-air, 44
  conventional, 42–43
  water-source, 44
  water-to-air, 44–46
Heat radiation, 7
Heat reclamation strategies, 37–40
Heat recovery strategies:
  from lighting systems, 40–42
  from refrigeration systems, 42–48
Heat wheels, 38
Heating, ventilating, and air conditioning (see HVAC equipment)
High sulfur coal, 201
HVAC (heating, ventilating, and air conditioning) equipment, 11, 24–28, 37, 69–70, 82, 135, 140
  checklist for, 239–242
  and control-point combinations, 85–88
  control systems, 100
  efficiency and building energy use, 28–29
  energy storage application, 106
  non-load-selective-type HVAC systems, 14
Hydraulic (pneumatic) structures, 17–18, 21
Hydrogen fuel, 235
Hydronic systems for transferring heat from lighting fixtures, 40–42

IEA (International Energy Agency), 226
Induction motors, polyphase, 135–138
  induction motor-generators, 137–138
  squirrel-cage motors, 135–137
  wound-induction motors, 135
Industrial energy consumption forecast, 185–194, 206
  and conservation, 194–195, 242–244
Institute of Gas Technology, 222, 223
Interior designer, role of, 29–31

International Energy Agency (IEA), 226
Intraindustry fuel substitution, 157–160, 177, 194
Iron and steel industry, 162, 166–170

Koppers-Totzek system, 197

Lamp types, 52–56, 59
LaRocca, Gregory P., 69
Lighting, 31, 33, 40–42, 48–59
   conservation opportunities in, 48–51
     checklist for, 238–239
   heat recovery from, 40–42
     air-return fixtures, 40
     bleed-off systems, 40–41
     hydronic systems, 40–42
     plenum-return systems, 40–42
     total-return systems, 40
     water-cooled luminaires, 41–42
   lamp types, 52–56, 59
   light fixture maintenance, 58
   luminaires, 56–58
   task lighting, 48–51
Lighting control, automatic, 31–33
Liquified gas, 225
Load leveling, electrical demand, 78–79, 107
Load programmers, 60
Load shedding, electrical demand, 65, 79–80, 104, 112
Luminaires, 41–42, 56–58
Lurgi system, 197

Manufacturing Chemists Association, 174
Materials, construction (see Building materials)
Mechanical recompression, 184–185
Merchetti, Cesare, 212
Modular integrated utility systems (MIUS), 140–143
Motors (see Induction motors, polyphase)

National Bureau of Standards (NBS), U.S., 29
National Energy Act (1978), 129, 158, 224–231
National Petroleum Council, 225
Natural Gas Policy Act (1978), 229
Nature, designing with, 12
NBS (National Bureau of Standards), U.S., 29

Nitric oxide, 201
Non-load-selective-type HVAC systems, 14
Nuclear energy, 212, 223
Nuclear pollution, 205–206
   (See also Environmental regulations)

Occupational Safety and Health Agency (OSHA), 232
Office of Management and Budget, 233
Office of Technological Assessment, 234
Oil industry (see Petroleum-refining industry)
Oil shale, 223, 230
On-site energy systems (OSES), 115–116, 147–154
   economic analysis of, 125–128
   fuel availability vs. purchased power cost, 134
   history of, 116–117
   maintenance of, 149–152
   potential applications of, 118–120
   prime mover selection, 120–125
   reliability of, 147–149
   siting of, 152–154
   solar power systems in conjunction with, 135
   (See also Cogeneration systems)
OPEC nations, 221
OSES (see On-site energy systems)
OSHA (Occupational Safety and Health Agency), 232

Paper industry (see Pulp and paper industry)
Passive-active options in solar power, 6, 15–16
Peltier-effect diffusion stills (PEDS), 180–184
Petroleum-refining industry, 162, 176–177
Plastics, 2
Plenum-return systems for transferring heat from lighting fixtures, 40–42
Pneumatic (hydraulic) structures, 17–18, 21
Pollution, nuclear, 205–206
   (See also Environmental regulations)
Polyphase induction motors (see Induction motors, polyphase)
Power-factor correction, 60–61, 81
Power management systems, computerized, 69
Power systems, Rankine-cycle, 138–142
Power techniques, regenerative, 62

Powerplant and Industrial Fuel Use Act (1978), 229, 231
Prestressed columns, 18–19
Prime movers in OSES, 120–125
Productivity, U.S., 234
Public Utilities Regulatory Policy Act (1978), 229, 231
Pulp and paper industry, 162–166

Quarles, John, 157

Rankine-cycle engines, 139–142
  with solar panels or collectors, 138, 140
  using organic fluids, 139
Rankine-cycle power systems, 138–142
  solar bottoming system, 141–142
  solar-powered, 138, 142
  solar and waste-heat bottoming system, 141–142
  using double-bundle condenser, 138
Rate structure (see Electric rate structure)
Reciprocating engines, 124–125
Recompression methods, 184–185
Refining, petroleum, 162, 176–177
Refrigeration systems, heat recovery from, 42–48
Regenerative power techniques, 62
Research and development (R&D), 232–234
Research Service, Congressional, 226
Rittenhouse, R. C., 115
Runaround system, 38–39

Scenarios in energy forecasting, 190–194
  economic scenarios, 191–194
  political scenarios, 191–194
Schlesinger, James, 219
Self-built structures, 17
Siting laws, 206–209
Solar collectors with Rankine cycle, 138, 140
Solar equipment industry, 320
Solar power:
  generation facilities, 107
  passive-active options, 6, 15–16
  systems in conjunction with OSES, 135
Spielvogel, Lawrence G., 23
Stanford Research Institute, 223
State air pollution laws, 230
State energy costs, comparative, 255–256
State environmental laws, 206–209

Steel industry (see Iron and steel industry)
Sulfur dioxide, 201, 202, 229
Sulfur oxides, 201–202
Sum loads, 6, 14, 16
Sun hoods, 14

Tar sands, 223
Task lighting, 48–51
Thermal energy storage (TES), 11, 15–16, 106–109
Thermal inertia of building materials, 9–11
Thermal lag, 8
Thermic diode, 15–16
Thermic panel, 15–16
Thermocycle economizer, 47–48
Thermorecompression, 184–185
"Total energy" systems, 117
Total heat wheels, 38
Trade deficit, U.S., 219, 227
Turbines, gas, 124, 131

U.S. balance of payments, 219, 227
U.S. Bureau of Mines, 222
U.S. coal reserves, 201, 223
U.S. Congressional Research Service, 226
U.S. Congress's Office of Technology Assessment, 234
U.S. Department of Commerce, 227
U.S. Department of Energy (DOE), 229–230
U.S. dollar, value of, 220
U.S. energy policy, 219–236
U.S. Environmental Protection Agency (see Environmental Protection Agency)
U.S. Federal Energy Administration (FEA), 29
U.S. Federal Energy Regulatory Commission, 231
U.S. Federal Power Commission, 206
U.S. Forest Service, 1
U.S. General Services Administration (GSA), 29
U.S. Geological Survey (USGS), 222
U.S. National Bureau of Standards (NBS), 29
U.S. Occupational Safety and Health Agency (OSHA), 232
U.S. Office of Management and Budget, 233
U.S. productivity, 234
U.S. trade deficit, 219, 227
Uranium, 223
Utility rates (see Electric rate structure)

Value-added principle, 160–162
Vapor recompression methods, 184–185
Voltage reduction, 58–59

Wasteful energy practices:
 in design, 19

Wasteful energy practices *(Cont.):*
 in lighting, 53
 in switching, 62
Weather, adverse effects from combustion process, 209, 211–213
Winkler system, 197